Schn Electric

Cassie Quaintance
Energy Segment Manager
North American Operating Division

1010 Airpark Center Drive
Nashville, TN 37217 USA
Direct: 615-844-8390 Cell: 303-882-5256
Fax: 859-817-4672

cassandra.quaintance@us.schneider-electric.com

Merlin Gerin
SQUARE D
Telemecanique

Marketing Green Buildings:
Guide for Engineering, Construction and Architecture

Marketing Green Buildings:

Guide for Engineering, Construction and Architecture

Jerry Yudelson, PE, MS, MBA, LEED® AP

Library of Congress Cataloging-in-Publication Data

Yudelson, Jerry.
Marketing green buildings: guide for engineering, construction, and architecture/Jerry Yudelson.
p. cm.
Includes bibliographical references and index.
ISBN 0-88173-528-0 (print) -- ISBN 0-88173-529-9 (electronic)
1. Green marketing. 2. Sustainable buildings--Marketing. 3. Architectural services marketing. 4. Architects and builders--Certification. 5. Construction industry. I. Title.
HF5413.Y83 2006
690.068'8--dc22

2005058224

Marketing green buildings: guide for engineering, construction, and architecture/by Jerry Yudelson

Published by The Fairmont Press, Inc.
700 Indian Trail
Lilburn, GA 30047
tel: 770-925-9388; fax: 770-381-9865
http://www.fairmontpress.com

Distributed by Taylor & Francis Ltd.
6000 Broken Sound Parkway NW, Suite 300
Boca Raton, FL 33487, USA
E-mail: orders@crcpress.com

Distributed by Taylor & Francis Ltd.
23-25 Blades Court
Deodar Road
London SW15 2NU, UK
E-mail: uk.tandf@thomsonpublishingservices.co.uk

Printed in the United States of America
10 9 8 7 6 5 4 3 2 1

0-88173-528-0 (The Fairmont Press, Inc.)
0-8493-9381-7 (Taylor & Francis Ltd.)

Table of Contents

Acknowledgements

My wife, Jessica Stuart Yudelson, (and my Scottie, *Madhu*) put up with the "magnificent obsession" to write this book, during evenings and weekends of 2004 and 2005 in Portland. This book is dedicated to them. Bob Fox of Cook+Fox Architects in New York generously updated me in September 2004 about the One Bryant Park development. Susan Kaplan of Battery Park City Authority, New York City, took the time to give me a tour of *The Solaire*, the first LEED® Gold apartment building.

My former colleagues at Green Building Services, Portland, provided some helpful information from their work on LEED-certified projects, as did my friends at Swinerton Builders, Portland. Thanks to Hamilton Hazelhurst, Vulcan Development, for his time in discussing how a major developer tackles green development, and to Alison Jeffries, marketing director, for her information on specific projects. I also benefited from talking with Jim Goldman and Rod Wille of Turner Construction about the implementation of their company's green building program.

We used some materials, with permission, from the April 2004 issue of *The Marketer*, the monthly magazine of the Society for Marketing Professional Services (SMPS). Karen Childress of WCI Communities in Florida guided me through the decision-making process of a major residential developer. The Civano community in Tucson shared some of their photographs, and my friends at Gerding/Edlen Development in Portland shared a copy of a 2004 advertisement for their green development projects.

My many friends and colleagues in the green building design, development and construction business in Portland, Seattle, San Francisco, Vancouver and other places, have always been there with helpful advice and solid information. Thanks also to the reviewers who lent their expertise, and to Lynn Parker for professional graphic design assistance.

How to Read This Book

People have different interests in the business of green buildings and sustainable design. This short guide will help you find the information you need. This book is meant to be read from front to back, but can be read one chapter at a time also. The tables, charts and graphs can also be read separately to gather useful (and often hard-to-find) information.

MARKETING ARCHITECTURAL SERVICES
Read Chapters 6, 7, 9 and 10 particularly.

MARKETING ENGINEERING SERVICES
Read chapters 6, 7, 9, 10, 13, 14 and 17.

MARKETING CONSTRUCTION SERVICES
Read chapters 6, 7, 9, 10 and 13.

MARKETING GREEN DEVELOPMENTS
Read Chapters 10, 11, 13 and 15.

UNDERSTANDING DIFFERENTIATION AS A GREEN BUILDING AND FIRM MARKETING STRATEGY
Read chapters 6, 8 (part three), 9, 10, 11 and 12.

MARKETING SOLAR POWER SYSTEMS
Read chapters 8 and 10.

MARKETING ENERGY EFFICIENCY TO THE COMMERCIAL BUILDING MARKET
Read chapters 5 and 8 especially, as well as chapters 2, 3, 14 and 17.

UNDERSTANDING THE MARKET FOR GREEN BUILDINGS
Read chapters 3 and 4 especially, as well as chapters 6 and 7.

UNDERSTANDING THE VIEWPOINT OF THE CLIENT FOR GREEN BUILDINGS
Read chapters 5, 6, 7 11 and 12.

THE FUTURE OF GREEN BUILDINGS
Read the forecasts in chapter 4 as well as chapter 18.

FINDING RESOURCES FOR FURTHER STUDY
Read the footnotes and chapter references, as well as the appendix.

SETTING BUSINESS STRATEGY FOR YOUR FIRM
Read chapters 6, 9 and 10, as well as the forecasts in chapters 4 and 19.

UNDERSTANDING THE GROWTH AND MARKET ACCEPTANCE OF THE LEED GREEN BUILDING RATING SYSTEM
Read chapters 2, 3, 4 and 5, as well as chapters 8, 9 and 18.

Chapter 1

Introduction

Let's review what has happened in the green building marketplace, since the introduction of the LEED system in April of 2000:

- Membership in the U.S. Green Building Council, the primary industry association has increased from about 600 corporate members at the end of 2000 to more than 5,500 members at the end of August, 2005, representing tens of thousands of design and construction professionals, as well as public agencies, environmental groups, building owners, property managers and developers.

- The LEED green building rating system has certified more than 350 completed projects, as of December 2005.

- More than 2,400 projects were registered at the end of December 2005 for certification under LEED, representing 50 states and 13 foreign countries, including Canada, Spain, India and China. Thousands of other projects are using the LEED evaluation system without formally registering with LEED. (Nearly 30% of the LEED registered projects are in California, Oregon and Washington, making the West Coast the "hot spot" of national green building activity at this time.)

- Nearly 29,000 professionals have taken the all-day LEED Technical Review eorkshop covering the basics of the LEED system.

- More than 21,000 building industry professionals have passed a national exam and become "LEED Accredited Professionals."

- The U.S. Green Building Council's third annual *Greenbuild* conference and trade show in Portland, Oregon, in November 2004 conference in Portland drew nearly 8,000 people, and the 2005 show in Atlanta, Georgia, drew nearly 10,000 attendees.

By anyone's reckoning, LEED is the fastest growing voluntary program to affect the design and construction industry in many years. LEED registrations are expected to grow more than 30% per year through 2007 and more than 25% per year through 2010. Understanding LEED and how to use it effectively in marketing a design or construction firm has become more important in the past few years. As clients' knowledge of, and comfort with, the LEED system grows over time, they will increasingly demand that their designers and builders understand how to use the system and how to achieve LEED results with little or no up-front design or construction cost. In effect, LEED has "raised the bar" for all building industry professionals. Not being up to speed on LEED, not having successful LEED projects in one's portfolio, will put firms increasingly at a significant disadvantage in our hyper-competitive marketplace.

This book raises and attempts to answer several key questions: How is green building marketing similar to all other types of professional service marketing, and how is it remarkably different? What available tools and techniques from conventional marketing can we use to greater effect in marketing green buildings? What is in fact the size of the market for green buildings? How can we estimate the future growth of this market? Who are the winners thus far in green marketing? How should a firm position itself to succeed in this growing marketplace?

To quote Tom Watson, the modern marketing genius behind IBM, "Nothing happens until a sale is made." Green building architects and engineers need a firm grounding in marketing theory and contemporary marketing strategy and tactics to be effective in this rapidly changing marketplace. Conventional professional firm marketers and account executives need to understand what the green building client, customer or consumer really wants, to be more effective in presenting green design features, sustainable strategies and new products to this new type of buyer.

PURPOSE OF THIS BOOK

This book presents the special features of marketing green buildings. It is designed for "professionals," people such as yourself whose livelihood depends on successfully marketing design and construction services, building projects, developments with green features and systems to serve these projects. There are thousands of us out there, trying to transform the building industry into a more energy-efficient and environmentally responsible activity, and we're doing it one presentation, one meeting, one design, one project, one product at a time.

ORGANIZATION OF THE BOOK

This book is organized into 19 chapters and appendix.

- Chapters Two and Three start out by looking at the green building market as it exists today, including the size of the market and especially the projects and products that have been successful from the late 1990s through 2005.
- Chapter Four presents factors influencing green building demand as well as short-term forecasts through 2007.
- Chapter Five looks presents the business case for green buildings, including both economic and "non-economic" factors.
- Chapter Six reviews survey data from specific experiences of green-design marketing that have successfully positioned firms for growth in this area, as well as case studies.
- Chapter Seven examines a few specialized "vertical" building markets, to see where the green building business is today and where it is trending.
- Chapter Eight looks at the current state and future success of several green-building technologies, and it offers a special focus on marketing solar power systems.
- Chapter Nine looks at the marketing approaches of selected professional service firms and also addresses the impact of a firm's differentiation in this field on attracting good people.
- Chapter Ten reviews classical marketing strategies for emerging markets such as green building and looks at the theories of "diffusion of innovations" that have characterized many similar innovative marketing efforts around the world.

- Chapter Eleven deals with the practical issues of selling green buildings and green-building services.

- Chapter Twelve addresses specific issues with marketing sustainable real estate developments, including single-family residential, condos and office buildings.

- Chapter Thirteen presents a discussion of the evolution of engineering design from "post modern" to "sustainable."

- Chapter Fourteen discusses the marketing and professional services opportunities in the "LEED for Existing Buildings" (LEED-EB) rating system.

- Chapter Fifteen discusses the marketing and professional services opportunities in the "LEED for Commercial Interiors" (LEED-CI) rating system.

- Chapter Sixteen looks to the future, with new editions of the LEED green building rating system, new technologies and new points of focus for this emerging industry.

- Chapter Seventeen discusses the professional engineer's role in applying the Energy Star® rating system to buildings.

- Chapter Eighteen projects the growth of the green building industry for the rest of this decade, using proven techniques from the field of technology forecasting.

- Chapter Nineteen discusses the people problem in growing and operating professional services firms in a time of high mobility among professionals in the building industry.

- The Appendix briefly lists some resources that you may find valuable on a continuing basis, including magazines, books, web sites and list-serves.

VALUING GREEN BUILDINGS

Green Building Rating Systems

Green buildings today have a variety of rating and certification systems available, but in the United States, the *de facto* national rating system is the U.S. Green Building Council's "Leadership in Energy and Environmental Design," or LEED. In certain "vertical market" segments such as secondary schools, modified versions of LEED are being used, for example in California and Washington state, with the standards posted for California by the Collaborative for High Performance Schools (CHPS, 2004), standards which have also been adopted for schools by the State of Washington. Similar standards are evolving in the health care industry, as exemplified by the "Green Guidelines for HealthCare" (see Chapter 7). All of these ratings systems are "point-based" and focus primarily on the building itself and the environmental impacts of its construction and operations. In 2005, a new competitor to LEED in the commercial buildings market was launched by the Green Building Initiative (www.thegbi.org), called "Green Globes," a self-certifying, web-based checklist system, but it hasn't achieved much marketplace interest in its first year of operation and can't be considered at this time as a serious competitor to LEED.

A survey of 167 public building owners in mid-2005 by a construction industry consulting firm found that 51% were familiar with LEED and, of those, 56% planned to implement LEED for some future project.[1] And 60% of the total survey group is incorporating energy-efficient elements in their designs. In the education sector, 73% have implemented some form of energy-efficiency designs or improvements in the past 12 months, with 40% of those projects using the LEED standards.

Benefits of Ratings Systems

Green building rating systems provide some value in the marketplace today, primarily to institutional building owners and developers, such as federal, state and local government; schools and universities; nonprofit organizations; hospitals, libraries, etc. These represent nearly 65%, for example, of the first 2,100 buildings registered under the LEED 2.0 system during its first five-plus years of existence, through September 2005[2]. Many private, for-profit building owners have used LEED to evaluate their buildings and to implement policies for sustainability and corporate social responsibility, e.g., American Honda Motor Company's LEED Gold building in Gresham, Oregon, and the Toyota North American campus in Torrance, California. A small handful of private businesses have built similar buildings for the benefit of their employees and/or to secure life-cycle cost savings in operations.

The Value of Green Buildings

The value of buildings depends on the nature of ownership. For example, a major government agency may construct buildings with a 50-year (or more) life, whereas a property developer may simply construct buildings for immediate leasing and short-term sales potential. Each of these building owners is pursuing different measures of value, and the task for green building marketers is to recognize this state of affairs and to tailor their approaches to different owners accordingly. (Chapter 5 discusses the business case for green buildings in more detail; the ability to articulate this case is important for architects and engineers who want to convince their clients to "go green" and to reap the business benefits of their commitment to sustainable design.)

Since marketplace values can shift rapidly, depending on the state of the economy, vacancy rates for properties, interest rates, etc., it is difficult to ascribe exact values for various green building measures. For example, in today's low interest rate climate in the U.S., where interest rates are at historically low levels, it is easier to justify longer-term investments in energy and water savings, both for government agencies and private building owners; in other words, the acceptable "payback" can be as long as seven to ten years, or more.

Buildings also accrue value by having lower operating costs. In a low interest rate climate, the multiplier of annual savings to get incremental increases in building value may be as high as 14 (cap rate of 7%), whereas in higher interest rate environments, it can shrink to 10 (cap rate of 10%). So, the same projected annual savings in energy and water costs, or benefits of productivity increases, might be worth 40% more in a low-interest-rate economy.

Marketing benefits might also accrue to LEED-rated green buildings, if they become the standard measure of value for commercial and institutional construction. Such buildings might be easier to lease or rent fully, or they might command higher rents or lease rates. At this time, there is little marketplace evidence that this is the case (see Chapters 11 and 12). If it were easier to lease green buildings, then speculative developers might be very interested, because a fully leased building prior to construction is a very valuable commodity. The LEED for Core and

Shell (LEED-CS) standard, currently in a "beta test" or "pilot" version, aims to assist developers with a pre-construction certification to help facilitate early leasing activity; it expects to launch a full-fledged version in the spring of 2006.

SUMMARY

Throughout this document, I rely on solid data, current through September of 2005. Most of this information is publicly available from the U.S. Green Building Council, from papers at green building conferences, from trade magazines or is based on my own projections and extensions of these data. I have also conducted several proprietary surveys and a large number of personal interviews to round out the picture of green building given in this book.

I am relying on AEE members, readers and users of this information, and fellow green building professionals to dialog with me and each other about the ways we can bring about a successful transformation of the building industry, to one that produces what most people say they want from it: energy- and resource-efficient, environmentally sound, healthy, comfortable and productive places to live, work, learn, experiment and recreate.

Thanks for your interest in this book, and happy reading! I welcome any other feedback, directed to me at my personal e-mail address: jerry.yudelson@comcast.net, or via my personal web site, www.yudelson.net.

Jerry Yudelson, PE, MS, MBA, LEED AP
Portland, Oregon
October, 2005

Chapter 2

Today's Green Building Market

Who are the winners in today's green-building market? Which firms have developed clear game plans and achieved obvious successes in marketing green building services and green building projects? Among the large architectural firms, giant HOK (ranked 28th of the largest design firms in the United States, based on 2004 billings, according to *Engineering News Record-ENR*)[3] stands out for its early commitment of a group to green buildings, its sharing of resources with others similarly committed in the late 1990s and its authorship of one of the leading texts on green buildings. (The HOK Guidebook to Sustainable Design, by Sandra F. Mendler and William Odell, New York: Wiley, 2000.) Perkins+Will Architects, number 64 on the ENR list and one of the top large international design firms (as part of DAR Group), is notable for having more than 400 LEED Accredited Professionals, as is Gensler, number 31 on the ENR list.

Among smaller architectural firms of less than 200 employees, a number of regional firms stand out, including BNIM Architects in Kansas City, Missouri (see case study at end of this chapter); Mithūn architects in Seattle, Washington (see case study in Chapter 9); LPA Architects in Irvine, CA; EHDD Architects in San Francisco; SMWM Architects, San Francisco; and Overland Partners in San Antonio, Texas. Each of these firms is led by a principal committed to sustainable design, participated in some of the earliest green-building efforts of the late 1990s, and has stayed abreast of the green building industry by making an aggressive commitment to innovation in this area. Not all firms and all principals of these firms share this passion, but those who do have also been able to attract smart and dedicated project architects and designers to their firms to implement their visions.

Also worthy of mention is Fox + Fowle Architects of New York City (now FX Fowle), where principals Robert Fox and Bruce Fowle created the landmark green high-rise, Four Times Square, the New York Times building; the 2005 high-rise residential project, *The Helena,* and other major green projects in and around New York. A few years ago, Robert Fox formed a new firm, Cook+Fox (www.cookplusfox.com) to construct what would be the largest LEED Platinum project ever, the 2.1 million sq. ft. *One Bryant*

Park project in New York City, which broke ground in 2004 and expects completion and occupancy in 2008 (Chapter 12 case study).

In the engineering field, some large national and international firms, including Flack + Kurtz in New York (plus San Francisco, Seattle, London, Paris and Washington, with 350 employees, number 226 in the ENR list), ARUP in London/New York/Los Angeles (73 offices, 7000 employees, number 77), and to some degree Syska & Hennessy (New York and Los Angeles, plus 12 other domestic offices and 600 employees, number 104 in the ENR list) have been able to carve out a niche as the preferred engineers for major projects by major firms. Their size, relatively few offices and cost structure have also allowed a number of regional firms to flourish in serving the needs of sustainable design-oriented architects. In Canada, Keen Engineering (acquired in October 2004 by publicly traded Stantec, TSX:STN; NYSE:SXC), has carved out an enviable niche as the green engineering firm of choice; in the past few years, Keen has extended its reach to a growing number of projects in the United States, more than doubling in size since 2000, now with 12 offices in the U.S. and Canada. At the beginning of 2005, Keen Engineering (now merged with Stantec) had shown a greater commitment to the LEED process than any architectural or engineering firm, with about 163 LEED Accredited Professionals in a staff of about 273[4]. Table 2-1 shows the top 10 construction industry firms with LEED Accredited Professionals, as of July 2005.

There are some specialized consulting firms active in this industry, but they are all generally smaller than 50 people and have "co-evolved" with the rise of the green building movement. None of the really large consulting engineering or pure consulting firms appears yet to have taken much of an interest in the green-design business. Some of the noteworthy consulting firms are CTG Energetics in Irvine, California; Paladino & Associates in Seattle; Green Building Services, Portland– of which the author was a co-founder; "7 Group" in Pennsylvania—a federation of independent consultants; and Elements in Kansas City, Missouri, a spin-off of BNIM Architects that is gradually taking on its own identity.

What do all of these firms have in common? They are technical leaders in sustainable design. They have been early entrants into the field. They have the size, scope and—in some cases—prime location to be at the nexus of sustainable design developments. They have worked on some of the landmark projects in this emerging industry. They are attractive companies to work for and as a result have attracted good young talent—a

Table 2-1. LEED Accredited Professionals[5]

Firm	Total Staff	LEED APs	% of Total Staff
Perkins+Will	900	419	46.6%
Gensler	1,866	418	22.4
HOK	1,892	303	16.0
Rudolph and Sletten	983	125	12.7
DPR Construction	617	121	19.6
Fluor Corp.	N/A	121	N/A
Stantec	4,273	120	2.8
Cannon Design	646	105	16.3
HDR	738	102	13.8
SmithGroup	727	90	12.4

must in the intense and highly competitive architecture, engineering and construction industry. They excel at personal and firm public relations, and they have participated in a variety of industry forums and associations. We will explore many of these attributes in the course of discussing how firms should market to the green building industry.

Consider these facts: there are more than 21,000 LEED Accredited Professionals as of September 2005 and nearly 29,000 have participated in LEED training workshops. But only 2,400 LEED-registered projects are on record, and less than 400 of these projects have been certified to date. So, it's not surprising that green-building industry leaders have yet to emerge—firms with 10, 20 or even 30 LEED-Certified projects under their belt. Many of the larger firms have in fact done fine green building projects without going through LEED certification, and many smaller firms have consistently won the "Top 10" annual awards from the AIA Committee on the Environment, with or without LEED certification.

Since LEED is a relatively new certification, barely six years old and, since it can take a year or more post-construction for certification to be achieved, it's not surprising that few firms have yet to take a strong market lead in this industry.

One other factor is also important: in general, architecture, engineering and building construction is a regional and even local industry, with few national firms except on the construction side; by and large, it has been the small- and medium-sized firms, looking for a market edge and more likely to be influenced by a few passionate designers or business people, who have seized the initiative in green design. The larger architecture, engineering and construction firms, with superior technical resources and strong client relationships, are now playing "catch up," *a fact that will dominate the green building market in the next half-decade.* Smaller firms will obviously be able to compete, but they may have to lower their sights in general toward smaller projects with LEED goals. Occasionally small firms can win larger projects based on design competitions.

LEED will continue to evolve: its goal is to serve only the top 25% of all building projects (personal communication, Nigel Howard, Chief Technology Officer, U.S. Green Building Council), and the "bar" will keep getting raised higher as more projects meet the current standards for higher levels of certification. LEED version 3.0, expected in 2007 or 2008, will raise this bar dramatically with its focus on rationalizing the LEED system across all credit categories and through the entire life-cycle of a building, a campus or urban district.

SURVEY DATA ON GREEN BUILDINGS AND TRENDS

A July 2004 Internet-based survey of more than 700 building owners, developers, architects, contractors, engineers and consultants, commissioned by Turner Construction Company, the country's largest commercial building firm, provides revealing data about the state of the green building market.[6] Looking ahead three years, 93% of executives working with green buildings expect their workload of green building projects to increase, more than half expecting the load to rise substantially. Of those executives currently involved with green building projects, 88% have seen a rise in green building activity the past three years, and 40% say a substantial rise.

About 75% of executives at organizations involved with green buildings reported a higher return on investment from these buildings, vs. 47% among executives not involved with green buildings. (It's not clear from the survey what "hard" data these expected returns are based on, other than projected energy efficiency savings.)

More importantly, of executives involved with green buildings, 91% believed that such buildings lead to higher health and well being of building occupants, as did 78% of executives not involved with green buildings. In other words, the business case for green buildings is stronger when health and well being are considered, than it is with strictly economic return on investment criteria. This is likely because green buildings are associated in most people's minds with daylighting, views to the outdoors for everyone, and higher levels of indoor air quality, whereas most people are less aware of projected levels of energy and resource savings associated with green buildings.

Greater experience with green buildings leads to more positive views of their impact on health and well being. Of those executives involved with six or more green building projects, 65% had a positive view of their impact on these issues, against only 39% of executives involved with only one or two green building projects.

Given these positive views of green buildings, it is surprising that the largest obstacles of widespread adoption of green building approaches are perceived higher costs (70% of all respondents cited this issue), lack of awareness regarding benefits (63%) and lack of interest in life-cycle cost assessment (53%), owing to short-term budget considerations.

LEED PROJECT TRENDS IN 2004

Tables 2-2, 2-3 and 2-4 show the growth of LEED-registered projects between July of 2003 and August of 2005, including number and size of projects. From these data, we can deduce some clear trends.

Owner Type

Overall growth in LEED-registered project numbers from mid-year 2004 to mid-year 2005 was about 52%. From Table 2-2, we can see that the greatest growth in projects by owner type, among the major players, occurred in the nonprofit ownership sector, followed by the private sector; federal and state government projects grew slower than the average. Although 44% of total registered projects through the middle of 2005 were from the government sector, the growth rate of that sector was below average. Hence the percentage of for-profit LEED registered projects is increasing slightly, as is that of nonprofit projects. For-profit owners account for only 26% of the total number of projects, but about 35% of all LEED project area, as these projects tend to be about 50% larger on average than all the other projects.

Project Size

Examining the data in Table 2-3, we can see that for-profit companies tend to build the larger projects, at about 151,000 sq. ft. on the average (based on 579 registered projects), compared with an average of 100,000 sq. ft. for all other projects. The estimated construction cost of these projects would range from $16 to $22 million, at about $110 to $140 per sq. ft. Federal projects represent the next largest average project size, by owner type, at about 132,000 sq. ft. each (based on 188 projects). State government projects are about 115,000 sq. ft. average, while the nonprofit and local government sectors build the smallest projects on average, except those owned by individuals. This is somewhat logical, given that local governments and nonprofits tend to build museums, recreation and cultural centers, libraries, fire and police stations, animal care facilities, and similar projects of smaller size. By contrast, the for-profit and federal government sectors tend to build larger office buildings (average size 134,000 sq. ft.), laboratories (139,000 sq. ft. average), multi-use (111,000 sq. ft. average) and similar facilities.

Average size of LEED-registered projects has decreased about 14% from 2003 to 2005, perhaps reflecting the more rapid growth of nonprofit

Table 2-2. Growth of LEED Registered Projects by Owner Type, 2003-2004

(SOURCE: USGBC, SEPTEMBER 2005, JULY 20, 2004 AND JULY 31, 2003 TALLIES)

Owner Type	July 2003 Projects	July 2004 Projects	September 2005 Projects	% of Total Projects, 2005	% Growth, 2003-2004	% Growth, 2004-2005
For-profit Corporation	237	372	579	26	57%	56%
Local Government	227	345	494	23	52	43
Nonprofit Corporation	138	272	441	20	97	62
State Government	100	174	260	12	74	49
Federal Government	81	142	188	9	75	32
Other	51	109	179	8	115	64
Individual	7	14	36	2	100	157
Total # of Projects	**841**	**1428**	2177	100%	70%	52%

sector projects, which tend to be about 30% smaller than the average of other projects.

Using the data in Table 2-4 and examining projects by building type, we can draw some interesting conclusions. The largest category of LEED-registered buildings is multiple-use facilities, which might contain offices, parking and ground-floor retail, for example. These account for nearly 31% of all LEED projects, as of mid-2005. Government projects make up 38% of all projects, by area, and 44% of all projects, so they are in general being built at an average size of about 106,000 sq. ft. (in the $10 to $15 million range of construction cost).

Among the larger number of LEED-registered projects, the faster-growing building types are:

- Multiple-use
- K-12 education
- Retail
- Multi-unit residential
- Health care

Interestingly, the growth of commercial office projects, by project size, was only half the growth by number, reflecting a smaller project size of 134,000 sq. ft. for new registrants. The reason for this probably reflects the growth of smaller office buildings from the nonprofit and local government sectors, as well as perhaps smaller private-owner buildings. The average size of new private-sector projects registered under LEED in the past year was 151,000 sq. ft., showing that the private sector continues to build large projects; not all of them are commercial or corporate offices, but include large multi-family housing projects, laboratories, health care and industrial facilities.

BUSINESS INTEREST IN GREEN OR "HIGH-PERFORMANCE" BUILDINGS

Owners and developers of commercial and institutional buildings across North America are discovering that it's often possible to have "champagne on a beer budget" by building high-performance buildings on conventional budgets. Many developers, building owners and facility managers are advancing the state of the art in commercial buildings through new tools, techniques and creative use of financial and regulatory incentives.

Table 2-3. Growth of LEED Registered Projects by Area, 2003-2005

(SOURCE: USGBC, SEPTEMBER 2005, JULY 20, 2004 AND JULY 31, 2003 TALLIES)

Owner Type	July 2003 (000 sq.ft.)	July 2005 (000 sq.ft.)	Average Project Size (000) Sq.Ft., 2004	Average Project Size (000) Sq.Ft., 2005	Percent of Total Project Area, 2005	Percentage Growth, Project Area, 2004-2005 (15 mos.)
For-profit Corporation	37,399	87,697	157.4	151.4	35%	49.9%
Local Government	24,381	45,237	94.3	91.6	18	39.0
Nonprofit Corporation	14,583	35,574	90.8	80.7	14	44.0
State Government	16,397	29,827	134.6	114.7	12	27.4
Federal Government	12,666	24,817	152.7	132.0	10	14.3
Other	5,938	21,791	138.5	121.7	9	44.4
Individual	410	2,547	55.4	70.8	1	228.0
Total Project Area	**111,774**	**247,493**				**40.0%**
Average Project Size (000 sq.ft)	**132.9**		**123.8**	**113.7**	**100.0%**	**-8.2%**

For the past 10 years, in ever increasing numbers, we have begun to see development of commercial structures for owner-built, built-to-suit and speculative purposes, using green-building techniques and technologies.

Understanding Green Buildings

What are people talking about when they speak of "Green" buildings or "high-performance" buildings? Typically, such buildings are measured against "code" buildings, in other words, structures that qualify for a building permit, but don't go beyond the minimum requirements. Additionally, such buildings are often measured according to a system such as the Advanced Building™ guidelines (www.poweryourdesign.com), the LEED (Leadership in Energy and Environmental Design) green building rating system of the U.S. Green Building Council (www.usgbc.org), the Collaborative for High Performance Schools (CHPS) ratings (www.chps.net), or in some cases local utility or city guidelines (a number of utilities have rating systems for residential buildings, for example). Also, such buildings typically have to "score" some minimum number of points above the "code" threshold to qualify for a "green" or "certified" or "high-performance" rating.

In six years, since the introduction of LEED in the spring of 2000, it has become for all practical purposes the "de facto" U.S. national standard. LEED is primarily a performance standard, in other words, it generally allows one to choose how to meet certain benchmark numbers—saving 20% on energy use vs. code, for example—without requiring specific measures. In this way, LEED is a flexible tool for new construction or major renovations in almost all commercial buildings across North America. There is a Canadian version of LEED that is almost identical to the U.S. version[7]; at this time, there is no Mexican version. LEED has proven its value as an aid to design teams tasked with creating green buildings.

As of September 2005, LEED had captured about 3% of the total new building market, with nearly 2,200 "registered" projects encompassing more than 247 million sq. ft. of new and renovated space. Currently, about 35 to 45 new projects per month are registering for evaluation under the LEED for New Construction (LEED-NC) system and 12 to 15 are being certified at this time. Since a project only gets "certified" under the LEED-NC system once it is completed and ready for occupancy, many projects are just coming up to the finish line of completing the documentation for a LEED rating. LEED provides for four levels of certification: "plain vanilla" Certified, Silver, Gold and Platinum. In 2003 and 2004, three projects in southern California achieved the Platinum rating; however, all three were projects for nonprofit organizations or government agencies. One was for a local utility, one was for a county park with the Audubon Society and one was for the Natural Resources Defense Council. Currently (September 2005), the largest Platinum project is a headquarters building in Boston for Genzyme Corporation, about 360,000 sq. ft. As of the end of 2004, nearly 170 projects had completed the certification process under LEED-NC. We project 153 new buildings will be certified in 2005 (Table 4-5).

To LEED or Lead?

What are the differences between using the other organizations' guidelines and using the LEED certification process? In one sense, they are complementary: using other guidelines can typically take a project more than halfway toward LEED certification. However, LEED focuses on a broader range of issues than most other green building or energy efficiency

Table 2-4. Growth of LEED Registered Projects by Building Type, 2003-2005

(SOURCE: USGBC, SEPTEMBER 2005, JULY 20, 2004 AND JULY 31, 2003 TALLIES)

Building Type	July 2003	September 2005	Percent of Total Projects, 2005	Average Size (000) Sq.Ft., 2005	Percentage Growth, by Number, 2004-2005
Multi-Use	160	672	30.8%	111	95%
Commercial Office	151	318	14.6	134	42
Higher Education	84	155	7.1	79	34
K-12 Education	52	133	6.1	121	60
Public Order/Safety	49	104	4.8	96	46
Multi-unit residential	32	97	4.4	147	52
Interpretive Center	45	77	3.5	28	18
Library	33	76	3.5	49	43
Industrial	33	71	3.2	140	29
Laboratory	27	52	2.4	140	13
Health Care	19	45	2.1	276	55
Assembly	14	31	1.4	169	35
Recreation	11	30	1.4	43	25
Finance & Communications	9	25	1.1	24	56
Retail	4	19	0.9	72	72
Military Base	3	17	0.8	57	42
Transportation stations	9	16	0.7	310	45
Animal Care	7	12	0.5	41	9
All Other	99	235	10.8	N/A	39%
Total Projects	841	2185	100.0%	113	53%

guidelines. For example, if owners' points of focus are primarily on energy use, reducing carbon dioxide emissions (linked to global warming) and improving indoor air quality, then a variety of Advanced Building guidelines can take them there efficiently. These improvements lead to reducing operating costs and improved occupant health, productivity and comfort. Both LEED and other building evaluation systems encourage an "integrated design" process, in which the building engineers (mechanical, electrical, structural and lighting) are brought into the design process with the architectural and interiors team at an early stage, often during programming and conceptual design. Integrated design explores, for example, building orientation, massing and materials choices as critical issues in energy use and indoor air quality, and attempts to influence these decisions before the basic architectural design is fully developed.[8]

For example, the "E-Benchmark" tool (*www.poweryourdesign.com*) of the Advanced Building guidelines from the New Buildings Institute (www.newbuildings.org) brings together more than 30 criteria for building designers to define and implement high-performance in building envelope, lighting, HVAC, power systems and controls. Each of these elements is critical in determining building performance, and they often interact in surprising ways. Through its development process, the E-Benchmark tool has considered these interactions and developed ways to incorporate them into some relatively simple tools for designers. In addition, the E-Benchmark tool covers every phase of the design and construction process, from pre-design charrettes to post-occupancy performance evaluation, forming a usable guide for designers to get from "cradle to graduate school," without spending huge amounts on research. The developers of this tool have documented energy savings of 20% to 27% in 15 major climatic regions of the United States, using sophisticated modeling techniques, for energy-conservation and energy-efficiency investments that have a three-year payback or less.[9]

What is the usefulness of these other guidelines? One current weakness of LEED applications for commercial projects is that only about 26% of all LEED "registered" projects (registration is the first step in the process, like getting engaged to be married) are in the private, for-profit segment of the market (currently, that translates to about 144-168 new registered projects over the past year, or about 12-14 per month in the entire country). Most LEED projects are in the institutional, public and nonprofit sector. Even fewer LEED projects are in the speculative commercial sector. That said, the type of LEED registered projects in the commercial market ranges from small suburban office buildings (15,000 sq. ft. or less) to very large financial and corporate buildings housing more than 500,000 sq. ft. of space. Because not all projects with sustainability goals decide to pursue LEED certification or actually follow through with the initial LEED registration, it is useful for designers to have other tools to ensure that their buildings are energy-efficient, without having to spend $15,000-$25,000 or more on energy modeling studies.

Given these weaknesses in the LEED application process for commercial and institutional buildings, it is important for some building developers and owners to have another tool for design that can be put into place immediately, either in conjunction with LEED or as a "stand alone" integrated design tool, such as E-Benchmark, so that "best in class" high-performance buildings can be built by the design professionals building owners and developers are most comfortable using for their projects. In addition, the E-Benchmark tool provides a designer with detailed guidance for the 15 major climatic regions of the U.S., from dry to humid and hot to cold; in this sense it is more detailed and "prescriptive" than the LEED performance standard.

THE PROCESS FOR CREATING A GREEN BUILDING

Often, the traditional "design-bid-build" process of project delivery works against the development of green buildings. In this process, there is often a sequential "handoff" between the architect and the building engineers, so that there is a limited "feedback loop" arising from the engineering aspects of building operating costs and comfort considerations back to basic building design features. In a more traditional design process, for example, the mechanical engineer is often insulated from the architect's building envelope design considerations, yet that set of decisions is often critical in determining the size (and cost) of the HVAC plant, which can often consume up to 20% of a building's cost. Also, the traditional "value engineering" exercise, held typically after it's obvious the project is over budget, often involves reducing the value of the HVAC systems by specifying lower efficiency (cheaper) equipment, possibly reducing the R-value of glazing and insulation, measures that will reduce first costs, but require the project to incur higher operating costs for energy for the lifetime of the building. (Lifetime operating costs are typically 80% or more of a building's total costs.)

As a result, key design decisions are often made without considering long-term operations. As we said earlier, most developers and designers find that a better process for creating green buildings involves an "integrated design" effort in which all key players work together from the beginning. Developers and owners have realized cost savings of 1% to 3% in building design and construction through the use of integrated design approaches as well as other "nontraditional" measures, which might include bringing in the general contractor and key subcontractors earlier in the process to help with pricing alternative approaches to achieve required comfort levels in a building.

Integrated design often involves "charrettes"—intensive design exercises—with key stakeholders during programming or conceptual design, as well as an "eco-charrette" with key design team members at the outset of schematic design. These charrettes are often an economical and fast way to explore design options as a group and all at once, before settling on a preferred direction. In the charrettes, everyone gets to provide input on building design before design direction is "set in stone." The owner or developer often gets to hear competing approaches to providing the space required and can be a more informed participant in the design process. For a good description of how this dialog might work, see articles by architect and LEED co-developer William Reed, AIA, on integrated design and regenerative design.[10]

A more effective refinement of the charrette process requires spending time on goal-setting sessions with the owner or developer and key stakeholders in the building process. These goal-setting sessions need to happen early on, and sometimes can take a full day to reach consensus. However, they often provide clearer guidance to design teams about preferred sustainability measures for the project and can assist in making budget-driven tradeoffs later in design.

Integrated design requires (considerably) more upfront effort, including dialog, charrettes, studies, timely decision-making, and so on, before the traditional start of a project with the schematic design phase. This implies that architects and engineers are going to require additional fees, and owners and developers are going to have to pay them, to get the results each party desires. On small projects, these fees might add 1% to 2% to the total project cost (1% of a $5 million project is $50,000, a typical amount for a full charrette-based design process with energy and daylight modeling studies, for examples), but perhaps pay for themselves in a quicker design process and possibly reduced HVAC system sizing, for example. (See a further discussion of these issues in Chapter 8, Part Two.)

Chapter 3

Industry Growth to Date

UNDERSTANDING THE "DIFFUSION OF INNOVATIONS"

To approach the green building market, it's useful to think of as a technological innovation. In classical marketing theory, people have found that such innovations take time to get into the marketplace. Typically, the time for more than 90% of the market to adopt an innovation is 15 to 25 years, i.e., a generation. In order to be adopted, an innovation typically has to have a major cost or business advantage over existing methods. In the author's experience, this advantage has to be greater than 25%, if cost alone is the criterion. This "cost-effectiveness barrier" exists because of the costs of learning new methods, the economic risk of investing capital to create new things, and the business risk inherently involved with trying something new. In the building industry, there has been historic resistance to discontinuous innovation, so that in many ways, buildings are built much the same as 20 years ago, relying on incremental innovations to improve performance.

Figure 3-1 illustrates how innovation enters the marketplace. Initially, a group of "innovators" with strong technical expertise and a tolerance for risk try something new. When the size of this group reaches about 2.5% of the total potential market, then a group of "early adopters" begins to find out about what the innovators are doing, observes successful field trials and then begins to incorporate the innovation into their own work. This group of "early adopters" has less tolerance for risk, but is attracted to the benefits of the innovation. When the size of the group adopting the innovation reaches about 16% of the potential market, then a new group, the "early majority," begins to use the innovation and begins the process of "mainstreaming" it. Finally, at about half the potential market size, a group of "late adopters" signs on, not wanting to be left out forever. At the end of the process, a group of "laggards" reluctantly adopts the innovation, and some people, of course, never adopt. (Think of the Amish, still driving a horse and buggy.)

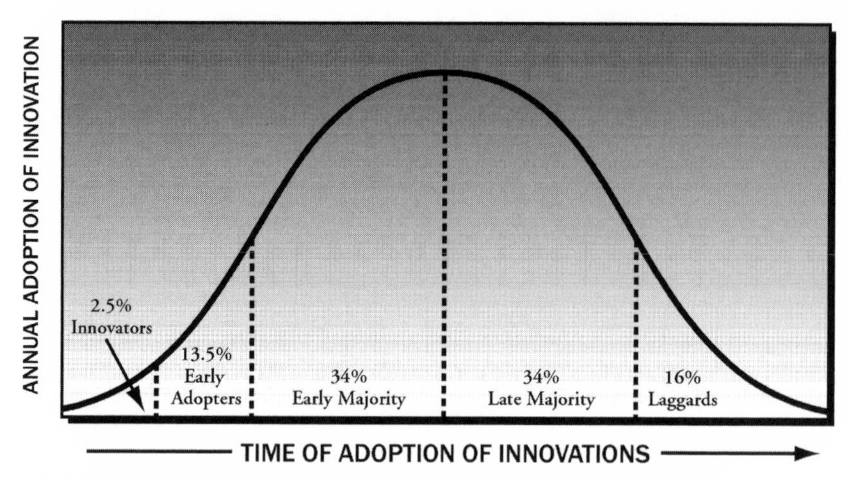

B) ANNUAL ADOPTION RATES AS PERCENTAGE OF TOTAL POTENTIAL MARKET.

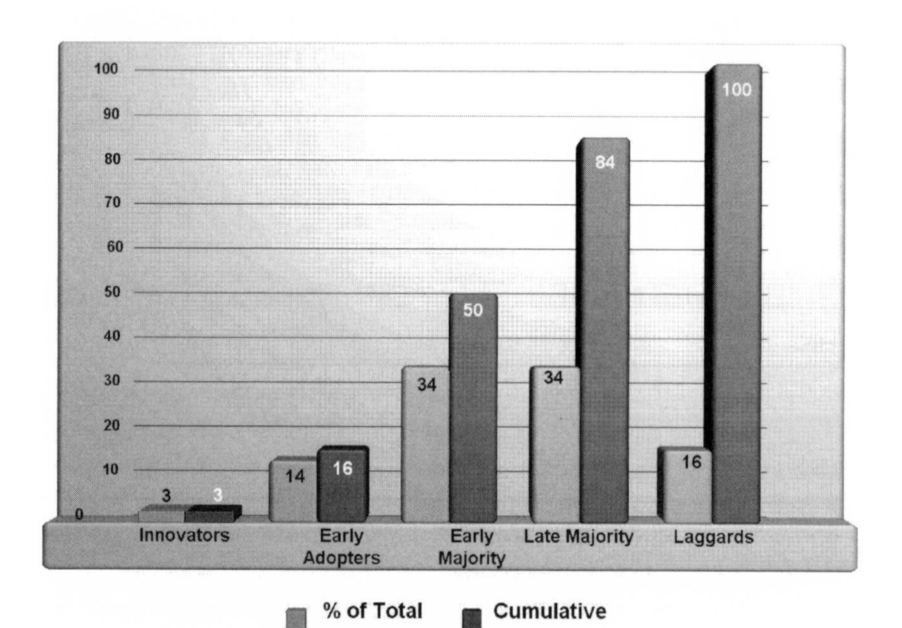

A) CUMULATIVE ADOPTION BY PSYCHOGRAPHIC TYPE

Figure 3-1. Diffusion of Innovation, Showing Total Adoption Rates by Phychographic Type

Of course, many technical and technological innovations never achieve mainstream status, owing often to cost or complexity. The process of mainstreaming is never smooth, and according to author Geoffrey Moore, it can be compared to "crossing a chasm" (see discussion in Chapter 10). Many technological innovations never have appealed to more than the early majority, either because something better comes along, or because they have high switching costs, offer few comparative economic benefits or are just too complex for the average user. One can think for example, of all the PDA products developed before the Palm Pilot™ finally came along and captured the mainstream business market.

GREEN BUILDINGS AS AN INNOVATIVE PRODUCT

To the degree that green buildings are simply "higher performing" buildings, one can argue that there's not much new here, that designing and building better buildings can readily be accomplished by the existing industry. However, if one considers the innovation to be rating and certifying buildings against various energy and environmental design criteria, as in the LEED green building rating system, then we can apply the classical theory of diffusion of innovation to forecast market demand. This theory encompasses the substitution of new ways of doing things for old ways, in a predictable pattern.

In addition, if we look at particular green building features that are becoming popular, then we could also apply this theory to forecast their adoption rates. In particular, one could look at the following technologies and forecast their likely individual market adoption rates, but that is beyond the scope of this book, at this time. We should note that certain products still have a lot of market skepticism, owing to concerns about longevity, maintenance costs and possible unintended consequences; such building technologies as green roofs, agrifiber MDF, waterless urinals and on-site sewage treatment certainly fall into this category.

- Photovoltaics (both stand-alone and building-integrated)
- Green roofs, for both aesthetics and stormwater management purposes
- Rainwater recovery and reuse systems, along with stormwater management systems

- On-site energy production, including wind and cogeneration systems, and fuel cells
- Water conservation products, including waterless urinals, low-flush toilets, etc.
- LEED-compliant roofs, including Energy Star® roofs that are high emissivity
- Low-VOC paints, sealants, coatings and adhesives

Cumulative adoption rates will follow some version of Figure 3-1, depending on how economically beneficial the innovation turns out to be. Each of the innovations listed above faces challenges to its adoption based on conventional economics, technical performance in the field, relative ease of specification, introduction by established competitors in the building industry, government and business mandates for change, and financial incentives from the government and utility sectors. These variables are shown in Table 3-1.

Table 3-1. Variables Determining the Rate of Adoptions of Innovations (after Rogers, 1995)

1. Perceived Attributes of Innovation	Examples: Relative (economic) advantage; compatibility with existing systems; complexity; trial-ability at reasonable cost; observable to others who might try it out
2. Type of Decision Required	Examples: Optional; group or committee decision; made by authority figure
3. Communications Channels Available	Examples: Mass media; interpersonal; web sites
4. Nature of the social system	Examples: Openness to innovation; network inter-connectedness to communicate results; changing norms favoring sustainability
5. Extent of change agents' promotional efforts	Examples: Writings, speeches, personal appeals

In Figure 3-2, the effect of a critical mass on the rate of adoption is shown graphically. According to Rogers (1995, page 314):

"The critical mass occurs at the point at which enough individuals have adopted an innovation so that the innovation's further rate of adoption becomes self-sustaining... An interactive innovation is of little use to an adopting individual unless the individuals with whom the adopter wishes to communicate also adopt. Thus a critical mass of individuals must adopt an [interactive communication] technology before it has utility for the average individual in the system."

While this example deals explicitly with communications technologies such as telephones, faxes, PDAs, teleconferencing and the like, it has clear relevance for green buildings. Given the large numbers of people now trained in the LEED system (more than 21,000 LEED Accredited Professionals, and about 29,000 who have attended the LEED training workshop), one can argue that LEED has all the hallmarks of a self-sustaining innovation. Therefore, its adoption rate can be predicted by utilizing this classical theory of innovation diffusion.

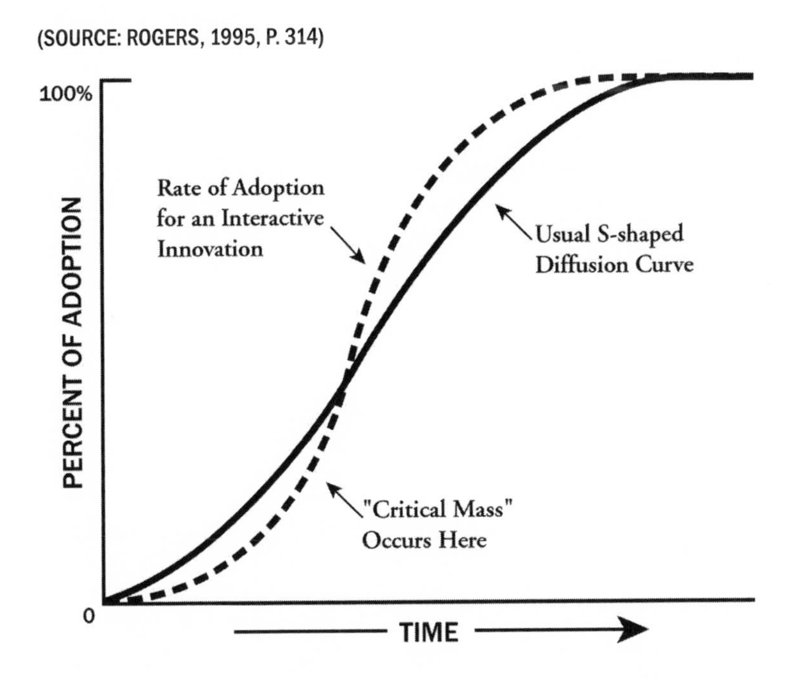

Figure 3-2. The Rate of Adoption for Innovations, Showing Effect of Critical Mass

According to Rogers, the critical mass occurs at the point at which enough people have adopted an innovation so that its further adoption is self-sustaining. Green buildings may represent a similar phenomenon, given the vast interconnected industry of designers, specifiers, builders, product suppliers and equipment vendors. In some cases, the "supply chain" for certain products such as certified wood may be under-developed in various regions of the country, hindering the desire of architects to specify it into their building projects, because of a lack of a "critical mass" of suppliers and contractors familiar with buying it.

MARKET HISTORY OF LEED BUILDINGS

The first step in a market forecast for LEED buildings is to see how the process has developed over the past five years. The current LEED system was introduced in April of 2000 (LEED version 2.0 and updates), following a pilot project in 1998 and 1999 to evaluate the proposed rating system.

As of September 2005, 285 projects had been formally certified and nearly 2,200 had registered with the USGBC as seeking eventual LEED certification (source: USGBC Member Update, September 2005). These numbers represent an increase since the end of 2003 of 80 certified projects and 667 registered projects. Projecting the 2005 activity levels (15 certified projects per month and 45 new registered projects per month) to the end of 2005 yields a projected total of 320 certified projects and 2,300 registered projects. Table 3-2 shows these data from the end of 2000 through the end of 2005, and Figure 3-3 illustrates the cumulative growth rates of registered and certified projects, as well as LEED registered project area.

We can draw a few conclusions from these data: the average LEED registered project at the end of 2003 contained about 130,000 sq. ft., and 121,000 sq. ft. at the end of 2004. As of September, 2005, that size had decreased to about 111,000 sq. ft. By the end of 2005, we expect the average project size to diminish to about 108,000 sq. ft. (The median registered project size may well be below 100,000 sq. ft., but there are no available data to verify this possibility.) The 2005 reduction in average registered project size indicates that smaller projects have begun to figure out how to participate in the LEED system and manage the fixed costs of meeting LEED prerequisites, such as commissioning and energy modeling, as well as the costs of preparing the documentation required for certification (or

Table 3-2. Actual LEED registered and certified projects, year-end 2000-2004 and 2005 year to date

(U.S. GREEN BUILDING COUNCIL DATA

Year Ending December 31st	New LEED registrations, version 2.x	Cumulative LEED v. 2.x registrations	Cum. LEED v. 2.x Project Sq.ft. (M)	New LEED certifications	Cumulative LEED certifications
2000	45	45	8.4	1	1
2001	230	275	51	4	5
2002	345	620	80	33	38
2003	457	1077	144	44	82
2004	715	1792	217	85	167
2005 (9 mos.)	369	2185	247	118	285
2005 (Proj.)	508	2300	260	153	320

Note: USGBC published data on LEED registrations and certifications are somewhat inconsistent, changing from new registrations to cumulative registrations, for example, or including (then excluding) pilot projects, such as those for Existing Buildings and Commercial Interiors. Therefore, these numbers may be revised somewhat in the future.

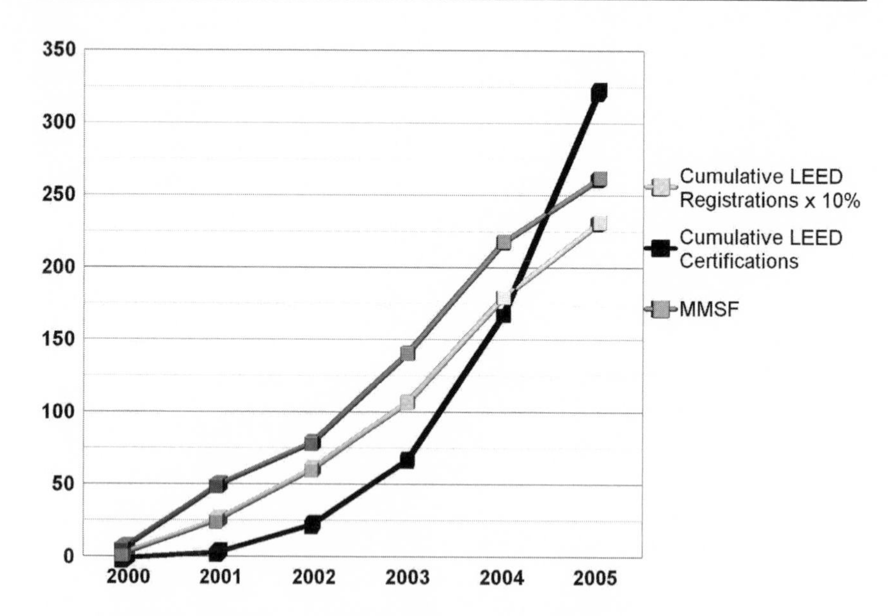

Note: first year of data is for 2000, sixth year is for 2005 (projected year-end data.) Observe the congruence of the growth curves, including LEED registrations (x 10%) and total LEED project area, in millions of gross sq. ft. However, note that LEED project certifications are now growing faster than new project registrations, partly as they "catch up" to the surge of registrations in 2003 and 2004.

Figure 3-3. Green Building Activity, 2000-2005

they may be registering with LEED but not completing the certification process because of these costs). Such smaller projects may also be more institutional, driven by policy considerations more than incremental cost.

The compound annual growth rate of LEED registrations for the first five years is about 100%. Over the past two years, the simple growth rate is about 114% for cumulative LEED registrations and over 80% for registered project area. These are phenomenal growth rates, no matter how the data are analyzed.

LEED Accredited Professionals and Workshop Attendees

Looking at another metric, the number of people trained in the LEED system, as of September, 2005 (USGBC *Member Update*), about 29,000 people had taken a LEED workshop, and more than 21,000 of these were "LEED Accredited Professionals (LEED APs)," having passed a national exam in the LEED system. With less than 2,200 registered projects to date, it's clear from the numbers that most of the LEED APs have yet to participate in their first LEED project. One might look on this as a factor for explosive market growth, or one could say that there's a lot more interest in LEED from the standpoint of design professionals than there is on the "building owner" or "developer" side of the industry.

LEED MARKET SEGMENTS

There are three easy ways to segment LEED registered projects, in terms of market impact:

- Geography
- Project Type
- Owner Type

Geography

Geographically, the top 10 states for LEED project registrations, in order of their total number, are the following (source: USGBC, as of September 2005):

California (344 projects)
Pennsylvania (119)
Washington (118)
New York (115)
Oregon (102)
Texas (88)
Michigan (82)
Illinois (79)
Massachusetts (76)
Virginia (65)
Arizona (65)

On a per-capita basis, small states such as Oregon and Washington lead the way, Oregon with about three times the national average project registrations per capita, and Washington with about 2.7 times the average (California—by contrast—is just 1.25 times the national average in registered projects per capita, surprising given that state's strong environmental advocacy on many other issues.) Including British Columbia project registrations in the total would make it clear that the West Coast has more than 25% of all LEED project registrations, but only about 16% of the total U.S. and Canadian population. Other areas of interest include the Great Lakes area, Texas and the New England/New York/Mid-Atlantic area. For marketing purposes, forecasts of LEED-registered buildings therefore should be very region-specific in the future.

Project Type

LEED registrations by project type are a bit harder to discern, because USGBC data groups many projects into a "multiple use" category, in which a primary use (for example, office buildings) might get classified as mixed-use because it also has other uses. With this caveat, the project types with the largest number of LEED Registrations are the following (data as of September 2005 from the USGBC Member Update newsletter), excluding multiple use projects.

1. Commercial office (21%)
2. Schools and colleges (19%)
3. Public Order/Safety (7%)
4. Multi-family residential (6%)

From this analysis, we can see that LEED is now widespread in the office building project-type category and fairly widespread in schools and colleges. Then there are a great number of mostly public projects that represent the next level of activity (assembly, interpretive center, library and public order/safety such as police and fire stations, as well as courthouses). Obviously, the public sector also represents much of the office building project type as well. Two interesting areas for potential future growth are multi-family residential, in which marketing advantages are gradually appearing, and the industrial category, likely driven by corporate sustainability objectives and policies.

Owner Type

In this segment, the classifications are easier to understand, possibly because they are fewer in number and more readily classified by type. Based on September 2005 USGBC data, Table 3-3 shows the owner types that are most prevalent in the LEED registrations.

Government Markets

From this information, it's clear that government and institutional users (including education and health care) have dominated the first five years of LEED project registrations, with 64% of the total registrations and 54% of the total registered project area (excluding "other/individual" project registrations for which no owner type is specified). Indeed, government-owned projects represent close to half (44%) of all LEED

Table 3-3. LEED Registrations by Owner Type, September 2005

Sector	Percentage of Total Project Registrations	Percentage of Total Project Area
For-profit sector	26	35
Local government	23	18
Nonprofit sector	20	14
State government	12	12
Federal government	9	10
Other/Individual	10	10

registered projects to date, indicating the prevalence of two driving forces in the green building marketplace: long-term ownership and operations perspective, and environmental policy considerations. (These two considerations are also likely driving the nonprofit and corporate sector LEED registrations).

What does this mean for marketers? One implication is that an intense focus on government and institutional projects is probably warranted at this time, since they are likely to have strong policies driving their use of the LEED rating system for project evaluation. A second implication is that larger private sector companies that are more likely to have sustainability or environmental stewardship policies and aspirations are also potentially valid targets; in the author's experience, however, the facilities and corporate real estate groups are often divorced from larger corporate goals and primarily concerned with lowering real estate costs for building projects, making the green building "sell" a bit harder to accomplish if it raises initial costs.

Higher Education Markets

According to LEED statistics, higher education project registrations made up about 8% of all LEED registered projects, numbering more than 155 in total, about 7% of the total registrations. Assuming there are about 3,000 colleges in the United States, starting right now an average of one project per year (3,000 in total) that would make LEED projects less than 3% of the college and university market (since these registrations have occurred over the past three years, they only represent about 80 to 100 per year). As of September 2005, there were only 44 certified higher education projects in the entire country (18% of the 246 projects listed as certified on the USGBC web site under LEED v. 2.x), so market penetration in this sector is just beginning. In the campus environment, surveyed by the author, at least 50% of the LEED projects exist because of support from the top leadership at the institution.[11]

Private Sector Markets

In the 26% of LEED-registered projects in the private sector, there are widely varying ownership types and perspectives. Many of the initial LEED projects have come from large corporations who have strong environmental stewardship goals and values and who have wanted to "walk the talk" in their (typically large) building projects. These include Ford, Toyota and Honda in the auto sector, The Gap, Goldman Sachs, PNC Bank, etc. In addition, there are many small business owners (including architects designing their own facilities) who have strong core environmental values that they want to illustrate in their own, typically smaller projects. Finally, there are a few speculative developers who have decided that LEED is the right thing to do and who have found that LEED goals and registration can confer marketing advantages; one such is the five-block, 1,700,000-sq. ft. "Brewery Blocks" commercial and residential project in Portland, profiled in *USA Today* on March 31st of 2004 (See case study in Chapter 12). In that project, the key people at the developer had very clear goals to make their project a significant statement that "green" can lead to "gold" in more ways than one!

LEED REQUIREMENTS FOR PROFESSIONAL SERVICES

When we speak of the green building industry and marketing to and for this industry, it's instructive to ask what's different or new about this industry, compared with the conventional practice of architecture, engineering and construction. The most obvious new thing is the role of LEED certification in defining what constitutes a green building. The role of certification obviously brings about a need for experts in the certification process, and so a small consulting industry has grown up around the need for LEED certification expertise. As of mid-2005, there were five architect-engineer teams around the country who review LEED certification documentation for the U.S. Green Building Council, and there will probably be 10 teams within a year, as the volume of LEED documents and projects seeking certification grows, and as new LEED products (LEED-EB, LEED-CI and LEED-CS) are added with different certification requirements. In total, a rough estimate would be about 100 to 200 people around the country who make all or most of their income consulting on LEED documentation. There are probably an equal number buried in the larger architecture/engineering firms who spend all or most of their time on LEED projects. Add several dozen consultants who provide facilitation and coaching services, and perhaps the total is perhaps 250 to 500 LEED consultants, serving about 2,200 LEED-registered projects and those projects with sustainability goals. This total might double in the next two to three years, but is unlikely to grow much larger, owing to the growing expertise of the 21,000 LEED Accredited Professionals in architecture, engineering, construction and related firms.

The real employment growth might be in energy modeling and commissioning, two technical areas in which a lot of mechanical engineers and technicians are rapidly being trained. Figuring 2,200 LEED registered projects with $20,000 energy modeling requirements, gives a total of about $40 million in professional services for this task. Assuming a two-year window for this modeling effort, there would be a requirement for about $20 million per year in modeling services, employing roughly 120 more full-time professionals (assuming billings of $170,000 per full-time modeler per year), for LEED modeling services alone.

The area of building commissioning offers significant potential for growth due to LEED requirements (in this case, it is a prerequisite). Assuming an average project size of 110,000 sq. ft. and a commissioning cost of $0.60 per sq. ft., and 200 projects per year requiring final certification, near-term employment from LEED-required building commissioning would likely result in $13 million per year in fees, employing about 100 additional professionals, at $130,000 average billings per commissioning professional.[12]

Other professional design and consulting services, less prevalent than energy modeling and commissioning, are likely to result in net employment of more than 200 people, as in-house LEED experts, lighting design labs and other nonprofit building industry assistance services, local USGBC chapter staff, materials experts, researchers, writing and publishing, etc. In many cases, these people already are employed and are just upgrading their skills.

The larger growth in employment is likely to come from the use of new materials and technologies in buildings, such as solar electric panels (photovoltaics, or PV), roof gardens, more energy-efficient glazing, rainwater reclamation system components, certified wood products, agrifiber products, products with low VOC emissions, products with higher recycled content and so on. Since most buildings are only going to add at most $2.00 per sq. ft. in capital costs, we can also calculate the employment impact of these expenditures at $220,000 per average LEED certified building (110,000 sq. ft. x $2.00), for 200 certified buildings per year, a total of $44 million annually, or up to $100 million if one assumes that project certifications will increase up to 500 per year (assuming no change in the average capital cost increase per LEED building, owing to experience in meeting LEED goals without increased costs). In manufacturing and distribution, average revenues per employee are likely to be $100,000 per year or more, giving the potential national employment benefit from the green transformation of the building industry of about 500 to 1,000 people. (See Table 8-2 for some higher green product estimates.)

One can assume that buildings are going to be of higher quality rather than generating large gains in employment. This is analogous to what has happened in the automobile industry over the past 10 years, in which new car selling prices, adjusted for inflation, have fallen, while the quality of the car and the equipment inside has increased dramatically. *The role of marketing, then, even in a growing industry such as green buildings, is to ensure that a firm does not lose market share to more innovative competitors and that the firm builds the internal capability to respond to industry changes and to increasing client requirements for green features and an integrated design process.*

To substantially increase employment in this industry, it would take a massive increase in federal and state funding, utility incentives and other grants for green building projects. This funding increase would likely need to be in the billions of dollars annually. The area it is most likely to happen is in new tax credits for energy efficiency measures and photovoltaics for buildings, as found in the new federal Energy Policy Act of 2005 (EPACT). For example, the State of Oregon is committed to spending for example up to $45 million per year in targeted energy-efficiency incentives from ratepayer funds and perhaps another $500,000 to $1 million in state tax credits for LEED-Silver (or better) projects ($100,000 incentive per project for 5 to 10 certified projects per year). However, few other states do this, outside of New York provides a state tax credit for LEED buildings.[13]

Chapter 4

Forecasting Demand for Green Buildings

Later in this chapter, we estimate the demand for LEED registrations and certifications, on a numerical basis, for the next three years. What in fact are the bases for understanding demand for green buildings, which might include LEED projects, projects that use LEED but don't formally register or certify, projects using the Advanced Building™ guidelines (e-Benchmark tool) or Energy Star™, or residential projects that use NAHB guidelines or other local utility certifications? These might include buildings with aggressive energy conservation goals, buildings that focus on demonstrating solar or other on-site power production technologies, building with green roofs, buildings with high-recycled-content products, and so on.

There are three basic approaches to customer/client analysis in determining the demand for green buildings. They are:

SEGMENTATION

Who are the biggest potential clients/customers? The most profitable? Logical groups based on needs, motivations and characteristics? Some variables could include: geographic location, price sensitivity, type of organization (public vs. private vs. non-profit), purpose or business of the organization (commercial offices, corporate user, higher education, K12 education, healthcare, recreation, cultural, etc.), number of buildings constructed each year, benefits sought, etc.

CUSTOMER/CLIENT MOTIVATIONS

What elements of green buildings do customers or clients value the most? Is it total cost of ownership, prestige, higher productivity of the workforce, response to a higher level mission statement or corporate purpose, or satisfaction of stakeholder demands? What are they really buying? What are the different motivations in each segment? How are these motivations changing?

UNMET NEEDS

What about green buildings that provide satisfaction of needs not being met with conventional building techniques? What problems are customers having with existing buildings that green buildings will address? Are customers really aware of their unmet needs, or do they have to be pointed out to them? Do these unmet needs represent leverage points for competitors?

By analyzing the potential customer base in this way, we can more fruitfully research who is buying, why they are buying or not buying now, and what they're likely to buy in the future. Knowing how customer demand for green buildings arises helps the marketer to understand how to present the concept to each particular audience. In the case of motivations and unmet needs, it is helpful to engage the client or customer in a dialog to identify exactly how, in their minds, a green building will satisfy some critical demand. Without this early research and dialog phase to establish why this green building project or product has value to the client or customer, and how much value, what usually happens is that the cost issue becomes predominant. The author has been part of numerous "green building" projects that foundered on this very issue of cost, because the architect or design team had not established value for the green building aspects early on in the client's mind, so when the time came to determine priorities for spending the limited project budget, the "optional" green elements were the first to go.

EXTERNAL FACTORS AFFECTING DEMAND FOR GREEN BUILDINGS

What external changes might affect customer demand for green buildings in the next few years? First of all, if the primary issue is cost, and secondarily risk of trying new things, we should be looking to establish the cost of green building projects, design measures and products as clearly as possible. The more green building projects are fully documented and reported on and the more design teams develop successful experience, the easier it will be to sell the next project because the perceived risk is less. So, what is changing in the external environment?

Growing Green Building Expertise

With more than 285 LEED certified projects and nearly 2,200 project registrations as of September 2005, there is clearly a lot more information in the marketplace. In addition, there are more than 21,000 LEED Accredited Professionals (by exam, as of September 2005), and more than 28,000 people have taken the LEED Intermediate Workshop, so that there is a growing body of expertise that projects can tap, all over the country. We are seeing many architecture firms and some engineering firms where 20% of the staff, even to 50% or more, has become LEED Accredited Professionals. Recently, we encountered a 130-person architecture firm in California where all seven principals are LEED Accredited Professionals, leading one to believe that this firm sees a potential competitive advantage in pushing each project to become a LEED-registered and eventually, LEED-Certified, building. So the development of professional expertise is clearly a positive factor driving LEED market growth in the past 12 to 18 months.

Cost Information

As more and more projects are certified, it is becoming easier to identify LEED-related and green building-related costs, making it easier to budget for such costs in the next project. It is also becoming cheaper to realize green building goals, especially LEED certification, as more building teams and consultants learn how to achieve these goals within conventional building budgets. Recent work by the Los Angeles office of the international cost consulting firm of Davis Langdon, offered evidence, based on 94 different building projects of vastly different types, that the *most important* determinant of project cost is NOT the level of LEED certification sought, but rather other more conventional issues such as the building program, type of design, the local construction economy, and other factors. In this study, the authors concluded that there was no statistically significant evidence that green buildings cost more per sq.ft. than conventional projects, primarily because so many factors influence the cost of any particular type of building.[14] If such analysis holds, there will be more pressure from owners and developers for design and construction teams to aim for high LEED goals, because these buildings are indeed perceived to offer higher value for the money spent. (See Turner

Construction Co. survey cited in Chapter 2).

An October, 2004 federally-funded study of the costs of achieving various levels of LEED certification for Federal government buildings draw somewhat opposite conclusions from the Davis Langdon study and supports somewhat similar conclusions to the work of Kats in 2003 for the state of California. In Kats' study of some 33 green building projects certified by the USGBC, the average cost of achieving various levels of LEED certification is shown in Table 4-1. For example, in this analysis, a $40 million public building seeking a LEED Gold level of certification might expect to budget about 2%, or $800,000, extra to achieve this rating.

Table 4-1. Incremental Capital Costs of 33 LEED Certified Projects (Kats, 2003)[15]

Level of LEED Certification	Average Green Cost Premium (% of total construction cost)
1. Certified (8 projects)	0.66%
2. Silver (18 projects)	2.11%
3. Gold (6 projects)	1.82%
4. Platinum (1 project)	6.50%
Average of 33 Buildings	1.84%

The study by Steven Winter Associates, released in October of 2004, carefully detailed for U.S. General Services Administration two typical projects, a new federal courthouse (with 262,000 sq.ft. and a construction cost of $220/gross sq.ft.) and an office building modification (with 307,000 sq.ft. and a construction cost of $130/gross sq.ft.), provided the analysis shown in Table 4-2. Basically, the incremental capital costs of LEED projects range from negligible for certified to 4% for Silver level and 8% for Gold level. It's worth noting that, faced with a potential 8% cost premium, many projects with a $40 million to $60 million construction cost budget would start looking at integrated design methods to lower a potential $3 to $5 million premium for LEED Gold.

Understanding the incremental costs of LEED certification efforts is important, as we shall see when looking a diffusion theory, because the single most important determinant of the rate of innovation adoption is the "relative advantage" of the innovation. In the construction world, construction costs are "hard," but benefits are mainly "soft," including projected energy and water savings, productivity gains, etc. Therefore, executing a cost-benefit analysis for each project is crucially important, to convince building owners and developers to proceed with the sustainable design measures and LEED certification effort.

Soft Costs

Soft costs for design and documentation services were also estimated in the GSA LEED Cost Study, and range from about $0.40 to $0.80 per sq.ft. (0.2% to 0.4%) for the courthouse and $0.35 to $0.70 per sq.ft. (0.3% to 0.6%) for the office building modernization project. One caution on these costs: they run from $100,000 to $200,000 for the courthouse and the office building modernization and may not be reduced much for smaller projects; therefore, the incremental percentage of total cost may be higher for smaller projects. Some typical soft cost elements and their ranges are shown in Table 4-3.

External Events

Outside forces have led building owners, buyers and developers to become increasingly concerned with long-term operating costs of their projects. These forces include the continual realization of the problem of global warming (through greenhouse gas—GHG—emissions), environmental hazards of mold, chemical allergies and other indoor air quality issues, lawsuits related to building mold and mildew from poor design and construction practices, current oil price escalations (reminding American consumers once again of their vulnerability to energy price increases), drought throughout the western United States, and many other factors. In the summer and fall of 2005, the rapid escalation in oil prices has had an impact on the psychology of consumers, building owners, developers and public officials, who are beginning to realize, for the first time in nearly 25 years, that energy prices are likely to be high for the foreseeable future.

In our opinion, the GHG issue of will become a key factor in driving major reductions in the energy use of buildings, including the incorporation of daylighting and natural ventilation approaches, so that it will not be unusual within three years or so, to see engineers aiming at 50% reductions in buildings' energy use (from current codes). An indication of the seriousness by which the business community views the issue of GHG

Table 4-2. Incremental Costs of LEED-Certifying Two Prototypical GSA Projects

(SWA, 2004)[16]

Level of LEED Certification	Range of Green Cost Premiums – (% of total construction cost)	
Building Type	New Courthouse	Office Modernization
1. Certified	-0.4% to 1.0%	1.4% to 2.1%
2. Silver	-0.0% to 4.4%	3.1% to 4.2%
3. Gold	1.4% to 8.1%	7.8% to 8.2%[17]

Table 4-3. Soft Costs forLEED Projects[18].

Element	Cost Range	Required in LEED?
1. Building Commissioning	$0.30 to $0.75 per sq.ft., $20,000 minimum	Yes
2. Energy Modeling	$12,000 to $30,000	Yes
3. LEED Documentation[19]	$15,000 to $30,000	Yes
4. Eco-charrettes	$ 7,500 to $30,000	No
5. Natural ventilation modeling	$ 7,500 to $20,000	No
6. Additional commissioning services	$ 3,000 to $15,000	No
7. Daylighting design modeling	$ 3,000 to $10,000	No (some utilities offer this as a free service)

emissions and associated global climate change is a September 2004 press release from The Conference Board, a major *Fortune 500* sounding board, in which its environmental expert stated:

> *Given the increasing costs of, and uncertainties surrounding, the reliability of traditional energy sources and growing pressures for higher standards of citizenship and contributions to global sustainability, businesses that ignore the debate over climate change do so at their peril.*[20]

Social and Cultural Changes

More and more stakeholder groups have become demanding and knowledgeable about green buildings, leading them to demand such projects for their schools and campuses, health care institutions, museums and libraries, and so on. This "grassroots" support is especially manifested in public and non-profit buildings, but will become increasingly evident, as support for, and understanding of, the concept of "sustainability" grows in the public's mind. On college campuses, sustainability is rapidly becoming a galvanizing issue for students and faculty alike, so that the push for LEED projects in higher education is likely to gain momentum.

Technological Changes

Many green building measures, such as underfloor air distribution systems, photovoltaics, rainwater harvesting, on-site waste treatment and green roofs, are becoming "mainstream" technologies that have a strong track record in design and use. As a result, they are developing a strong supportive infrastructure of salespeople and suppliers, a better cost history, understanding of how to bid and install them from contractors, and increasing advocates among architects and engineers who are learning to design and specify such systems. The construction industry infrastructure is quite mature and highly complex, and it's important that green building marketers master its intricacies in order to get new green building designs, technologies and products into that marketplace.

Economic Changes

Today's low-to-moderate interest rates, likely to persist for several more years, owing to high levels of productivity and worldwide supply overcapacity in many industries, have the effect of encouraging capital investments that yield long-term operating cost savings, because the "present value" of future savings is larger than in a higher interest rate environment. In addition, the relative lack of investment in energy supply infrastructure in recent years may have the effect of guaranteeing higher future energy prices. As a result, the "payback" of capital investments for energy and water conservation, for example, becomes more favorable with each passing year. It is fairly easy to justify, for example, a 10-year payback (return of initial investment in annual energy savings) for energy conservation and efficiency investments, at least on rational economic grounds, which could lead buildings to be built 30% to 50% more efficient than current energy codes require.

Political and Legal Changes

Many more cities and states are adopting incentive programs for green buildings, including direct financial incentives. This has led to direct investment by the private sector in such areas as Washington, Oregon, California, New York and British Columbia, to name just a few, to take advantage of such investments. In September 2005, Congress passed the Energy Policy Act of 2005 (see Table 4-4), which provided significant new tax credits for solar energy systems placed in service in 2006 and 2007; in addition, if the cost of oil and gas remain high, it is likely that these incentives will be extended beyond calendar year 2007. Regulations for these credits have not yet been written, but energy engineers and other design professionals should take advantage of them for projects that will be completed by the end of 2007.

Legal changes are also occurring, as liability for poor indoor air quality becomes an issue in lawsuits and claims against builders and developers, architects and engineers, contractors and specifiers. LEED and other certification programs provide some "risk management" or "damage control" benefits, by providing objective standards by which design professionals and their clients can argue that they were pursuing "best practices" in their projects and by requiring documentation of actual achievements.

Industry Practices

Just about every area of the country and every sector of the marketplace awards design contracts (and, more and more, "CM/GC" contracts to builders) based on qualifications, as opposed to fee or price. Fees are then negotiated after a selection is made. Therefore, it is becoming increasingly difficult for firms to qualify for a "short list" of finalists for a project, without having a strong green building orientation, knowledge

Table 4-4. Energy Policy Act of 2005 (EPACT) Key Provisions[21]

Affected Technology	Tax Credit
Photovoltaics	30% (residential limit is $2,000 credit)
Solar thermal systems	30% (residential limit is $2,000 credit)
Microturbines	10% (up to $200/kW credit)
Energy conservation investments for HVAC, envelope, lighting and water heating systems	$1.80/sq.ft. (tax deduction if exceeding 50% savings vs. ASHRAE 90.1-2001 standard)
New homes exceeding 50% savings vs. model code	$2,000 credit for site-built homes

of green building products, and some successful projects under their belt. Competitive pressures alone are driving more firms toward green building projects, even if they don't really "believe" in the need for them. It is also leading them to hire younger professionals who are advocates for green buildings and who form a positive influence inside the firm or company for taking this approach.

Certification Programs

LEED is not the only green building certification program that has an effect on green building demand. There are many certification programs developing to handle subsets of the LEED rating system that affect building products, indoor air quality, "green tags" for carbon dioxide emissions, "cool roofs", green roofs, and many similar measures. Also, state-level and utility programs serve the residential building market. We will also begin to see products rated for their impact on energy demand, greenhouse gas emissions, use of Persistent Bio-accumulative and Toxic (PBT) compounds and similar non-product features, all falling under the rubric of "Life Cycle Assessment." Users will increasingly be given evidence of a product's origins and full life-cycle impact. Specification writers will increasingly incorporate these product features and environmental characteristics in construction documents. As the evidence for a product's "green-ness" gets clearer, those that stand out can expect to be specified and used more and more.

FORECASTING DEMAND FOR LEED GREEN BUILDINGS

Given the latest full year's data (2004), indicating a slowing of the "explosive" growth rate of LEED's first four years, September 2005 updates from the USGBC, and a full-year projection for 2005, shown in Chapter 3, we estimate growth rates of 28% to 31% per year in cumulative LEED registrations for the next three years and derive the results shown in Table 4-5.

There are obviously some key issues in this forecast. For example, the biggest hindrance today in registering and certifying LEED buildings is the perceived (and often real) additional cost. This has been demonstrated in many surveys, including two proprietary surveys I have conducted. In assessing the diffusion of innovation as an operating principle for projecting demand for LEED buildings, *it is cost-effectiveness above all* that determines the rate of adoption of new technologies. As individual building owners and developers, along with design and construction teams, get more experience with LEED buildings and with sustainable design measures and technologies, we can expect the "cost premium" for LEED buildings to decrease, resulting in perhaps a substantial boost to these estimates.

SIZE OF THE NON-RESIDENTIAL BUILDING MARKET

LEED has always been aimed at just 25% of the building market at any given time. With that in mind, it is useful to estimate the total available market for LEED registrations. Table 4-6 shows data on U.S. non-residential building markets in July 2005. This analysis leaves out the relatively few LEED residential projects at the present time (even though residential construction represents 55% of the total U.S. construction market and 67%, or two-thirds, of total U.S. building construction). Of the non-residential building market, only 41% ($114/$276 billion in 2005) is public construction, which constitutes the largest LEED market, representing more than 70% of total project registrations.

LEED Market Share

Assuming that the average building cost is $100 per sq.ft. for new construction, that leaves a potentially available LEED market of about 690 million sq.ft. per year. With a split of 59% private and 41% public, the

Table 4-5. Estimated LEED Registrations, Project Area and Certifications, 2005-2007

Year Ending December 31st	New LEED project registrations	Cumulative LEED registrations	Cum. LEED Project Sq.ft. (million)	New LEED project certifications	Cumulative LEED certifications
2000	45	45	8.4	1	1
2001	230	275	51	4	5
2002	345	620	80	33	38
2003	457	1077	141	44	82
2004	715	1792	217	85	167
2005 (estimated)	508	2300	260	153	320
2006[2] (estimated)	700	3000	330	363	683
2007[2] (estimated)	840	3840	410	430	1113

private LEED market could be about 407 million sq.ft./year and the public LEED market would be potentially 283 million sq.ft./year.

Over the five-year period, 2001-2005, LEED expects to have registered about 260 million sq.ft. This total would represent about 182 million sq.ft. of public projects (at 70%) and 78 million sq.ft. of private projects. *During that five-year period, the public market would have been about 1,415 million sq.ft., so LEED's market share would be about 12.8% of the total available public market; the private market would have been about 2,035 million sq.ft., and LEED's market share would represent about 3.8% of the total available private market.* Overall, LEED's market share, assuming our 2005 projections (of 43 million new sq.ft. of project registrations) are realized, would be about 6.2% of the total available LEED market, or about 1.55% of the total U.S. non-residential building market (less if one removes the 8% of LEED multifamily residential projects).

Looking at the 690 million sq.ft. of new (potential LEED) non-residential building construction per year, and an average LEED project size of 110,000 sq.ft., this *gives a potential LEED market of 6,270 buildings per year*. If LEED is planning to register about 508 projects in 2005 (our projection, Tables 3-2 and 4-5), the market share for this year would be about 8.0% of the available market. *In other words, less than 10% of the potentially available LEED projects are expected to register with LEED in 2005. For a program that's five years old, that's still not a large market penetration.*

SUMMARY OF SHORT-TERM FORECASTS

Forecasting demand for LEED projects (Table 4-5) would result in about 840 new LEED project registrations in 2007, averaging about 95,000 sq.ft. (80 million sf total, about 3% of the total non-residential building market, or about 12% of the total available market for LEED projects, assuming the 25% number given above), with about 430 new projects receiving some level of LEED certification in that year. This level of LEED activity would represent a significant part of the building industry activity, and it will certainly have an influence on many other aspects of the industry. From a diffusion of innovation perspective, one can see that the green building movement will have moved to the "early majority" phase of the total available market, with much of what is now still considered innovative becoming commonplace.

Table 4-6. U.S. Non-residential Building Market, 2002-2005 (Billions of Dollars)[22]

Building Sector	July, 2004 (annualized rate)	May, 2004 (annualized rate)	2003	2002	Estimated Total LEED Market at 25%
Lodging	12.1	$12.4	$11.1	$11.0	$ 3.0
Office	44.7	44.4	41.4	46.0	$11.2
Commercial	70.3	63.1	62.3	63.4	$17.6
Health care	37.5	33.1	29.9	28.1	$ 9.4
Educational	75.8	75.1	74.2	71.9	$18.9
Religious	7.6	8.2	8.5	8.3	$ 1.9
Public safety	9.6	8.6	9.0	9.4	$ 2.4
Recreation	18.3	19.7	20.0	19.8	$ 4.6
Total	275.9	$264.6	$266.4	$257.9	$69.0

Chapter 5

The Business Case for Green Buildings

Deciding to design to green building guidelines is always a challenge, when budgets are so tight and schedules are so compressed. The developer or owner needs to have a really clear idea WHY this is so important. Let's recap the "business case" for green buildings[23]:

- **Reduced operating costs**. Green buildings will save on operating costs for energy for years to come; with the price of oil above $50 per barrel and the prospect of peak period electricity prices zooming up again, it just makes good sense to design the most energy-efficient building possible. Even with "triple net" leases in which the tenant pays all operating costs, it makes sense for landlords to offer tenants buildings with the lowest possible operating cost. Many green buildings are designed to use 25% to 40% less energy than required by current codes; some buildings even achieve higher levels of efficiency. Translated to a normal operating cost of $1.60 to $2.50 per sq.ft. for electricity (the most common "fuel" for buildings), this means a reduction of $0.40 to $1.00 per year in utility operating costs. Often these savings are achieved for an incremental investment of about $1.00 to $3.00 per sq.ft. With building costs ranging upwards of $100 per sq.ft., doesn't it make good sense to invest 1% of capital cost in securing long-term operating cost savings, particularly those with "paybacks" less than three years? Given the current constrained market for commercial rents in most regions of the country, these savings add to the bottom line quite quickly. In an 80,000 sq.ft. building, this translates into $32,000 to $80,000 per year into the owner's pocket.

- **Reduced maintenance costs and greater occupant comfort**. Energy saving buildings that are properly "commissioned" at costs of $0.50 to $1.00 per sq.ft. of initial cost (equal to one year of savings), show additional savings of 10% to 15% in energy costs, and are easier to operate and maintain.[24] By having a comprehensive functional testing of all energy-using systems prior to occupancy, it is often possible to have a smoother running building for years, because all those "little" problems get fixed before occupants start using it. Also, a key part of the commissioning process is documenting operator training. In this way, new personnel can be trained how to keep the facility running at peak efficiency. In non-commissioned buildings, the handoff between operators can be brief and often incomplete.

- **Improved productivity**. The service economy continues to grow. Productivity gains for healthier indoor spaces are worth anywhere from 1% to 5% of employee costs, or about $3.00 to $30.00 per sq.ft. of leasable or usable space, given average employee costs of $300 to $600 per sq.ft. per year (based on $60,000 average annual salary and benefits, and 100 sq.ft. to 200 sq.ft. per person). With energy costs typically under $2.50 per sq.ft. per year, it appears that productivity gains may easily equal or exceed the entire energy cost of operating a building! For corporate and institutional owner/occupiers of buildings, that's too much money left on the table to ignore.

- **Risk management**. "Sick Building Syndrome" lawsuits will likely continue in coming years. With the national focus on mold in buildings and its effect on people, developers and owners need to re-focus their attention on indoor air quality. Green building certifications can provide some measure of protection by having a third-party certification of measures installed to protect indoor air quality, beyond just "meeting code."

- **Stakeholder relations/occupant satisfaction**. Tenants and employees want to see a demonstrated concern for both their well-being and for that of the planet. Savvy developers and owners are beginning to realize how to market these benefits to a discerning and skeptical client and stakeholder base, using the advantages of green building certifications and other forms of documentation, including local utility and industry programs.

- **Environmental stewardship**. Being a "good neighbor" is not just for building users, but for the larger community as well. Developers and owners are beginning to see the marketing and public relations benefits (including branding) of a demonstrated concern for the environment. Many larger public and private organizations have well articulated "sustainability" mission statements and are beginning to see how their real estate choices can both reflect and advance those missions.

- **Increased building value.** Increased annual energy savings will also create higher building values. Imagine a building that saves $37,500 per year in energy costs versus a comparable building built to "code" (this might represent a savings of $0.50 per year per square foot, for a 75,000 sq.ft. building, for example). At a "capitalization rate" (effective discount rate) of 7.5%, this would add $500,000 (or nearly $7 per sq.ft.) to the value of the building! ($37,500/0.075 = $500,000). For a small upfront investment, an owner can reap benefits that typically offer a payback of three years or less, and an internal rate of return exceeding 20%, for what is *nearly a sure bet*: energy costs will continue to rise faster than the general rate of inflation and faster than rents can be raised.

- **More competitive product in the marketplace.** There is a dawning realization among speculative developers that green buildings can be more competitive in certain markets, *if* they can be built pretty much on a conventional budget. Whether for speculative or build-to-suit purposes, green buildings with lower operating costs and better indoor environmental quality should be more attractive to a growing group of corporate, public and individual buyers. "Greenness" will not replace known attributes such as price, location and conventional amenities, but green features will increasingly enter into tenants' decisions for leasable space and into buyer's decisions to purchase properties for the long haul.

HIGH-PERFORMANCE BUILDINGS

What are the design and operating distinctions of high-performance buildings? They save 25% to 30% or more of building energy use by incorporating high-efficiency systems and conservation measures in the basic building envelope, HVAC plant and lighting systems. These systems and efficiency measures can include extra insulation, high-quality glazing and solar control measures; Energy Star® appliances such as copiers, computer monitors and printers; building orientation and massing to utilize passive solar heating and cooling design; high-efficiency lighting (often using high-output T-5 lamps in many applications); carbon dioxide monitors that monitor room occupancy and adjust ventilation accordingly, so that energy is not wasted in ventilating unoccupied space; occupancy sensors—which turn off lights and equipment when rooms are unoccupied; and higher-efficiency HVAC systems, variable speed fans and motor drives, to produce the same comfort level with less input energy; and many similar techniques.

High-performance buildings are "commissioned," through the use of performance testing and verification before the end of construction, for all key energy-using and water-using systems. Typically, commissioning involves creating a plan for all systems to be tested, performing functional testing while the mechanical and controls contractors are still on the job, and providing the owner with a written report on the performance of all key systems and components. Green building commissioning involves third-party peer reviews during design, to see if design intent has actually been realized in the detailed construction documents. Finally, most commissioning programs also involve operator training and documentation of that training for future operators.

Systems such as the Advanced Building guidelines provide a clear definition of various levels of building commissioning, so that you can choose the approach that best suits your situation and get it priced for the actual services you require. Some programs also include an "end of warranty period" or "11-month" spot check on all key systems and preparation of a manual for "re-commissioning" the systems at some set interval, typically five years. Think of commissioning as analogous to the "sea trials" a ship undergoes before it is handed over to the eventual owners. No ship would be put into use without such trials, which may expose flaws in design or construction, and no building should commence operations without a full "shakedown cruise" of all the building systems that use energy and affect comfort, health and productivity. Often, the documentation provided by the commissioning process can be helpful later on in troubleshooting problems with building operations.

Such buildings achieve higher levels of indoor air quality through a careful choice of paints, sealants, adhesives, carpets, and coatings for the base building and tenant improvements, often in conjunction with building systems that provide higher levels of filtration and carbon dioxide monitors to regulate ventilation according to occupancy. With so many building occupants having breathing problems and chemical sensitivities, it just makes good business sense to provide a healthy building. Documentation of these measures can often help provide extra backup when fighting claims of "sick building syndrome." This benefit of "risk management" is an often overlooked aspect of green building guidelines, but can often be useful to demonstrate to prospective tenants or occupants the often "invisible" measures taken by building designers and contractors to provide a safe and healthy indoor environment.

Healthy buildings incorporate daylighting and views to the outdoors not only for occupant comfort, health and productivity gains, but also to reduce energy costs. There is a growing body of evidence that daylighting, operable windows and views to the outdoors can increase productivity from 5% to 15% and reduce illness, absenteeism and employee turnover for many companies. Throw in higher levels of building controls that allow for such things as carbon dioxide monitoring and ventilation adjustments, for example, and one has an effective program addressing the "people problem" that can be sold to prospective tenants and other stakeholders. For owner-occupied buildings, these savings alone are often enough to justify the extra costs of such projects. Considering that 60% or more of the operating costs of service companies (which most are) relate to employee salaries and benefits, it just makes good business sense to pay attention to productivity, comfort and health in building design and operations.

UNDERSTANDING "VALUE PROPOSITIONS"

Proponents of green buildings often resort to rhetoric ("green is good") when advancing their case. But building owners and developers have very different approaches to valuing green buildings. These "value propositions" can include both measurable and non-measurable benefits, both for the building itself and for the organization. Examples of measurable benefits include the life-cycle savings in energy and water consumption from improved building energy performance, as well as an improved market positioning resulting from building a green building; non-market measurable benefits include intangibles such as public relations benefits to an organization or the prestige of locating one's own business in a certified green building.

Green buildings or sustainable construction projects often involve more expense than conventional construction, especially in "soft costs" for additional design, analysis, engineering, energy modeling, commissioning and certification to relevant standards such as the U.S. Green Building Council's LEED program. These costs may exceed 1% of construction costs for large buildings and 5% of costs for small buildings, depending on the particular measures employed. Higher levels of sustainable building (e.g., the LEED "Silver" or "Gold" standard) may involve additional capital costs, based on case studies of buildings in the U.S., as documented in studies by the Rocky Mountain Institute (2002) and Kats, et al. (2003).

Justification of such additional costs has traditionally rested on the economic "payback" or return on investment for energy and water conservation measures. Green building standards such as LEED incorporate additional requirements beyond energy and water use, e.g., indoor environmental quality, use of recycled materials and sustainable site considerations, so it is increasingly difficult to justify green building investments on the value of utility savings alone.

Additional value propositions for green buildings are needed to justify the incremental expenses involved. We can classify these value propositions as occurring inside the building and outside the building. "Inside the building" values may be created by increased productivity of workers, while "outside the building" values may be created by enhanced image of a company or institution. Additional measures of value might include risk management reductions, improved recruitment and retention of key employees, and enhanced value of real estate investments.

COSTS OF GREEN BUILDINGS

Value is Relative to Cost

Each measure in a green building project has a cost, even if it's just the cost of documenting the LEED rating level. Such professional services, including energy modeling, building commissioning, additional design services and the documentation process, can easily add 0.5% to 1.5% to the project's cost, according to several studies. So, green-building proponents might have to find measures by which to value their projects to overcome the additional costs of such projects. Evidence is increasing that buildings certified at the LEED "Gold" level might add only 2% to 3% to project cost (Ecotrust, 2002; Kats, 2003, see Chapter 4), but might create total stakeholder values far in excess of these costs, measured in terms of "net present value" or enhanced image.

Costs Depend on Many Factors

Many of the green building measures that might give a building its greatest long-term value, e.g., on-site energy production, on-site stormwater management and water recycling, "green roofs," daylighting and natural ventilation, often require a higher capital cost. While it is possible to get a LEED-certified building at no additional cost, as one moves to make a building truly sustainable, cost increments often accrue. Finding out which costs are going to provide the highest value is a primary task of the architect, working in concert with her client, the building owner or developer, and the builder.

Classifying the Value of Green Buildings

The classification scheme in Table 5-1 shows how to understand and use green building value analysis. Almost all discussions today involve only the upper left quadrant. However, many owners and developers value other attributes of green buildings more highly than operating cost savings.

Measurable Benefits of Green Buildings

Table 5-2 shows a variety of measurable benefits; these occur both inside and outside the building, according to a classification scheme developed by the author.

Benefits that Accrue to the Building Itself

These costs are fairly easy to measure and include the usual energy and water savings of well-designed buildings, of which there are many examples, such as the 200 case studies documented by Rocky Mountain Institute (2002). Other benefits might include resale value, owing to such savings, as discussed above.

Benefits Related to Building Occupants and Their Behavior

There is a growing body of literature, documenting the very real and measurable benefits of buildings that have enhanced daylighting, natural ventilation and improved indoor air quality. These benefits are found in such areas of concern as enhanced productivity, reduced absenteeism and illness, and improved retail sales. A total of nearly 200 peer-reviewed studies in the academic literature attest to these benefits, according to Professor Vivian Loftness of Carnegie Mellon University. These studies are presented in the Building Investment Decision Support tool (BIDS),

Table 5-1. Understanding the Benefits of Green Buildings

Type of Benefits/Where They Occur	**Measurable**	**Non-measurable/Intangible**
Inside the Building		
Economic Non-economic	Energy/Water Savings Increased Employee Well-Being/ Productivity	Increase in Building Value Increase in Employee Morale
Outside the Building		
Economic Non-economic	Marketing and Sales Retention/Recruitment	Brand Identity Public Relations

Table 5-2. Examples of Measurable Economic Benefits Inside the Building

Type of Benefit/ Building Owner	**Energy Savings Investments**	**Water Savings Investments**	**Productivity Enhancements**	**Sales Enhancements**
Commercial				
Speculative	Justify only with higher rents	Justify only with higher rents	Must add to sales value	Not likely
Owned	Reduce operating costs	Reduce operating costs	Add daylighting & views outside	Add daylighting to retail
Non-commercial				
Owned Leased	Reduce operating costs	Reduce operating costs	Reduce no. of employees	Not applicable

developed by Loftness. (This is a proprietary database at present but some aspects can be seen at *http://cbpd.arc.cmu/ebids*.)

Benefits that Accrue to the Organization or Building Owner

Green buildings may also yield benefits to the building owner, for example, through higher rents, better tenants or longer-term leases from quality tenants. These immediate benefits might also translate into a higher resale value, since the resale value will typically be a multiple (the "cap rate") of the projected annual cash flow generated by the building. Additional benefits might include sizable tax credits for green buildings that are being offered by several states.

Reducing Costs for "Churn"

For long-term owner-operators such as government agencies and large corporations with open-plan offices, green buildings measures such as underfloor air distribution systems (raised or access floors, utilizing displacement ventilation techniques) may reduce costs of churn, the incessant propensity of such organizations to move people's work areas, typically at rates of 20% to 30% per year. Savings of up to $2,500 per workstation in moves have been reported, based on an average workstation area of 100 sq.ft. per person. If the underfloor system adds a net cost of $3 to $5 per sq.ft. to the initial cost of the project, then that cost is nearly recovered in the first year or two ($2,500 x 20% = $500; 100 sq.ft. x $5/sq.ft. = $500). Meanwhile, the benefits of reduced energy costs and healthier air accrue to the project from the beginning.

Non-Measurable or Intangible Benefits of Green Buildings

Benefits that Accrue to the Building Itself

Most intangible benefits accrue outside of the building operations. A few of these are shown in Table 5-3.

Benefits that Accrue to Building Occupants

Many employees may feel benefited by the enhanced prestige of working in a well-known building, or may have higher morale owing to the better physical and psychological environment. This result has always been the goal of architects, but only recently have there been the tools for analyzing and simulating daylighting and natural ventilation, for example.

Table 5-3. Importance of Typical Intangible Benefits Outside the Building

Type of Building/ Benefit	Public Relations	Public Policy	Marketing and Sales	Company or Organization Brand/Image
Commercial				
Speculative (developer)	Somewhat important	Not applicable	May help with loans or leases	Could be useful for developers
Corporate-owned	Very important	Not applicable	Very important	Very important
Non-commercial or Institutional				
Owned	Very important	Crucial	Not applicable	Very important
Leased	Very important	Crucial	Not applicable	Very important

Benefits that Accrue to the Organization or Building Owner

This area reflects many of the primary benefits of green buildings, and most of the benefits are intangible. Some of these are brand image, public relations, enhanced marketing capability, market positioning, reduced risk of lawsuits, employee loyalty and attractiveness to new employees, fund-raising capability, "doing the right thing," and the like. While these benefits are "intangible," they are nonetheless "real," in the sense that they do have economic or social value.

Brand Image

Large corporations are highly concerned with brand image. Recent green building projects by such large consumer products companies as Ford Motor, Honda America, Toyota (in southern California), The Gap (clothiers) and many regional and national banks, all serve to indicate the importance of brand image and the role that green buildings might play in enhancing it, by appealing to a customer base of "Cultural Creatives" (Ray and Anderson, 2000).

Public Relations

Many public agencies have sought to demonstrate their commitment to sustainability through the construction of green buildings, including the U.S. General Services Administration, which owns or operates more than 500 million sq.ft. of buildings. Other public entities with strong green building programs include the State of California and the City of Seattle, which have each committed more than $400 million to LEED "Silver" certified new building construction in the next few years, and the City of Vancouver, British Columbia, which recently announced a LEED Gold commitment for all public buildings greater than 5,000 sq.ft.

Enhanced Marketing Capability

One local residential remodeling company in Portland, Oregon, built the first LEED-certified building in the state, with a showroom for consumers. In the process, it has garnered considerable publicity for the company and has built a reputation as a place to go for certified wood products in cabinets, according to CEO Tom Kelly of Portland-based Neil Kelly Remodelers, one of the largest remodeling firms in the country. Other firms are moving to demonstrate their own sustainability commitments through their building and facilities management programs.

Market Positioning

The author has worked with a smaller (2,000-student) private university that is embarking on a green building program. During the course of a green building charrette in 2001, it became apparent that the university could "re-position" itself in the marketplace for students in its local area, from its current image as a place for teacher training to a much more progressive "sustainability" image, through both building design and a conscious effort to integrate sustainability into the curriculum. The economics of such a re-positioning are attractive: at the margin, each additional student generates a gross revenue of $80,000 to $100,000 over a four-year college program (assuming no tuition subsidies), with almost no marginal cost for serving that student—the university will hire no new instructors or build any new classrooms for a handful of new students. If enrollment were to grow just 0.5%, i.e., 10 new students, as a result of a LEED-rated building program, revenues would increase by more than $800,000. This revenue gain would justify the extra costs of a green building program.

As part of an overall sustainability program, green buildings may assist a company's market positioning with both consumers and other businesses. Companies that export products to more "green-friendly" regions such as Europe may find such positioning essential for success.

Reduced Risk of Lawsuits

In the litigious climate of the U.S., employers would do well to consider how a documented improvement, according to accepted "best practices," as exemplified in LEED and related standards, might serve as a defense in a lawsuit, for example, alleging "Sick Building Syndrome" as the cause for an employee's illness or harm. How much better a defense would this be than merely citing the building code as the rationale for design of building mechanical and control systems? In addition, it is possible that, over time, this reduced risk of lawsuits might allow insurance companies to offer lower rates for such buildings. Thus, a risk management approach (intangible benefit) might eventually result in tangible economic benefit.

Employee Loyalty and Attractiveness to New Employees

Many organizations seek to demonstrate their commitment to employee health and welfare through a variety of benefit programs. Companies and agencies are now beginning to view green buildings as a tangible and positive statement of their long-term commitment to employees' health and well-being. Employees can also be expected to see such benefits as "real," especially when the employer takes care to communicate what it is doing and how the building is better. One may also conjecture that the prospect of working in a well-known green building might, at the margin, also be a powerful attractant to high-level professional employees. At a current cost of $30,000 to $150,000 to recruit, hire and train a new high-level employee, this intangible benefit might well yield positive results. As another example, with explosive growth in population of secondary students now underway in the U.S., there is a growing teacher shortage. Would it be possible for school districts with new schools offering the best in daylighting, natural ventilation, controllability of classroom environment, etc., to attract better teachers, compared with those schools than can't offer such amenities?

Fundraising Capability

Consider the case of a small private university, cited above, with a limited base of charitable donors, typically limited to local business people and alumni. This university feels that a green building program can help it tap a new base of donors, not only for the buildings, but for the university's new "sustainability initiative." This need to "walk the talk," creates opportunities for green building advocates to help universities and nonprofits such as Ecotrust (von Hagen, et al., 2003) to build leading-edge projects, by working with them in the fundraising arena. An excellent tool would be a fundraising prospectus highlighting the benefits of the green building and the commit to sustainability it represents by the organization. (A commitment to LEED Platinum for an historic renovation project resulted in a $100,000 Kresge Foundation grant to a Portland nonprofit in 2004, to foster an integrated design process for the project).

"Doing the Right Thing"

The green building literature is replete with examples of projects moving ahead, because the owners or developers realize it's the "right thing" to do. Unabashed altruism is still present, even in the commercial building industry, and organizations that want to stay on the leading edge of change recognize increasingly that their building programs reflect on their character. In the oft-quoted words of the British statesman, Sir Winston Churchill, "We shape our buildings, and then our buildings shape us."

CASE STUDY: OWNER DECISIONS IN THE PORTLAND, OREGON, AREA

As described elsewhere in this book, Pacific Northwest green building activity leads the nation: LEED registered projects in Oregon and Washington (at the end of 2005) are about three times what one would expect based on population alone (See Table 10-1). What makes these green buildings valuable to their owners? Why are some projects willing to pay the extra costs of "going green"? How can we understand what values are embodied in green buildings?

In terms of *direct economic benefits,* many projects are using green building techniques to save energy and water costs (to a lesser degree) in future building operations. For example, architect Heinz Rudolf of BOORA Architects (www.boora.com) expressly set out to better the 2001 Oregon energy code by more than 40% in designing the new Clackamas High School, a 268,000 sq.ft. building for some 1,800 students. The new school, which opened in April 2002, expects to use 44% less energy than a conventional code-compliant high school, based on energy modeling done for the project. In this case, Rudolf had to achieve this result with no increase in the building budget. (Rudolf was assisted by Portland General Electric, Rocky Mountain Institute and other entities that paid for early modeling and collaborative design efforts.) The project received the LEED Silver certification. This project's green goals have been largely the result of the architect's strong commitment to sustainable design and his very long-term relationship of trust with key individuals at the school district.

Many green building projects aim at lowering utility bills by 30% or more, while staying within a conventional budget. This suggests that direct economic benefits of more than 30% utility savings are either very difficult to achieve or not highly valued enough to be a major design goal. As more buildings begin to show energy savings potential of 40% or higher, we can expect owners' expectations to rise and design teams to respond accordingly.

In 2006, a new project will open in Portland, The Center for Health & Healing for Oregon Health & Science University, Oregon's main teaching hospital. This building is expected to receive a LEED Platinum rating and, at 400,000 sq.ft., would be the largest such Platinum building. Designed by GBD Architects (www.gbd-architects.com) and Interface Engineering (www.ieice.com), both of Portland, the project has a "first-cost" SAVINGS of more than 10% of the original $30 million budget for mechanical and electrical components, with a projected 61% reduction in energy costs and 54% in total water use. This project demonstrates the benefits of an integrated design process and an experienced developer and design team willing to "push the envelope" of building design, in generating a high-performance building on a conventional building budget.

Indirect economic benefits are certainly being discussed widely in green building circles, particularly the higher productivity and improved health and morale of employees, but it is difficult to find projects that have explicitly incorporated those goals into the green building design program, except for the inclusion of daylighting (LEED standard met in 40% of the first 100 certified projects) and views of the outdoors (LEED standard met in 71% of the first 100 projects) as design elements. What we are finding is that most green building projects are putting considerable effort to adding and improving ventilation and indoor environmental quality (with such measures as low VOC paints, sealants, carpets, etc.), which they expect will result in improved morale and health. However, there are no projects that are actually measuring such results, and most projects are not yet spending a lot of effort to educate the building tenants about the health benefits of their new building.

Direct non-economic benefits seem to be the big driving force in many of the green building projects finished or underway in the Portland metro area. For example, the first two LEED-certified buildings in the Portland area, the 15,000 sq.ft. Viridian Place office building in Lake Oswego, Oregon[25] and the 70,000-sq.ft. Ecotrust Natural Capital Center in Portland, were very explicitly aimed at demonstrating the owners' commitments to sustainability and, in the case of Ecotrust, a nonprofit foundation, to garner national publicity to help in future fund-raising efforts.

In terms of *indirect non-economic benefits,* a good case in point is the Honda America facility in Gresham, Oregon, a 228,000-sq.ft. distribution, training, and office building for the American Honda Motor Co., and a LEED-Gold certified project that is intended to help demonstrate the company's global commitment to sustainability[26]. (However, one could argue that Honda's hybrid cars are in fact likely to create far more favorable publicity for Honda in the sustainability arena than any company facility.)

A similar driving force is the State of Oregon's *North Mall* office building in Salem, which is aiming at LEED certification as one means to satisfy Governor John Kitzhaber's executive order of 2000, which mandated sustainable actions for all state agencies. The North Mall building had not started design when the executive order was promulgated, so it was able to benefit from the push that the order gave to state agencies to do sustainable design. The state's project architect also had a role in guiding the North Mall building in that direction. The design team, led by architect Nels Hall of Portland's Yost Grube Hall Architecture (www.ygh.com) had a major role in helping the state to define how to incorporate sustainable principles into the design of this 110,000-sq.ft. building.

This desire for recognition as a sustainability pioneer has motivated Lewis & Clark College's new *Howard Center for the Social Sciences*, which is aiming at a LEED Gold rating, according to Michael Sestric, the college's campus planner, and project architect Will Dann of Thomas Hacker & Associates (www.thomashacker.com). The college was determined to have the new project LEED certified at a reasonably high level. Sestric feels that much of the college's recent building projects have been very environmentally sound and energy-efficient, but recognized that a LEED certification offered an independent measure of the college's efforts to build sustainably and served as a benchmark for continued improvement.

In the commercial arena, the *Brewery Blocks* project in Portland is moving toward LEED certification of its five major buildings (total of nearly 1.7 million sq.ft. of mixed-use commercial, retail and residential), not only to demonstrate the sustainability commitment of its developer, Gerding/Edlen Development (www.ge-dev.com), but also to compete effectively in attracting tenants in a soft real estate market. Dennis Wilde of Gerding/Edlen, long active in Portland sustainable design circles, has been the guiding hand at Brewery Blocks toward creating the best possible sustainable design. He has been very clear that the green goals of this major project must be met within a conventional budget. Wilde has also created a tenant- improvement handbook to guide future commercial building tenants toward more sustainable choices (see further discussion in Chapter 12).

It appears at this time in the Portland area that the prospect of LEED certification is a significant motivator for green design, especially in competition with other local buildings of the same type. This is primarily a *non-economic* benefit to the owners and developers of these projects, one that they hope will garner significant public relations exposure for their projects and also serve as a physical expression of their longer-range commitment to sustainable policies and projects. Will we begin to see more green projects built for explicit economic benefits, or will they continue to be designed and built for primarily non-economic reasons?

Green building advocates, mostly architects, building engineers and sustainability mavens, need to learn and use the language of business and marketing in order to be more effective at market transformation. This language includes of course, economics and finance, as exemplified in "return on investment," future value of buildings, productivity, etc. However, the non-economic and intangible language of business, found primarily in the areas of marketing, risk management and public relations, needs to be equally emphasized. In fact, it is often the driving force for key business decisions to build greener buildings at this time.

Chapter 6

Experiences of Green Marketing

In 2003, the author surveyed nearly 500 green building practitioners, using a database of attendees from the 2002 U.S. Green Building Council conference, *Greenbuild*, held in Austin, Texas. The survey results displayed a variety of methods used for marketing green building services, as well as some of the challenges faced in the marketplace.

To find out what it would take to accelerate green building adoption, we surveyed the marketplace of green practitioners for guidance. We also compared our survey results to a similar survey of 523 practitioners published by *Building Design & Construction* magazine in November 2004, as a "Progress Report on Sustainability."[27] Later in this chapter, we present the results of a similar survey of building industry professionals, conducted by Turner Construction in the summer of 2004.

SURVEY PREPARATION

Our survey of 2,700 people in July 2003 who attended the first U.S. Green Building Council annual conference and exposition in November 2002 in Austin, Texas, used a web-based survey tool from www.zoomerang.com and a 20-question survey instrument of our own devising. We got 473 responses or about 17% of the total population solicited.

SURVEY PARTICIPANTS

Survey participants came from a range of disciplines and occupations, including those shown in the table on the following page.

This distribution of respondents is very similar to that of a survey conducted in the Fall of 2003 and Fall of 2004 by a major trade magazine, *Building Design & Construction* (BD&C, 2003, 2004 Green Building White Paper), which showed, for example, 35% architects, 11% engineers and 9% government agency personnel, but which comprised a population with much less experience in sustainable design and green building projects

Category	Percentage of Total
Architect	35
Engineer	9
Other Design Team	5
Construction Team	11
Building owner/developer	6

than that represented by our survey. Where a similar question was asked, we have included the results of the BD&C survey along with ours. In terms of measurable experience, 73% of our respondents characterized their firms as "very experienced" or "somewhat experienced" in sustainable design, compared with 49% of BD&C survey participants in 2004.

SURVEY RESULTS

Participants were asked to characterize how their firm has responded to the market.

Firms that were more committed to sustainable design had a tendency to try to LEED-certify a project, create specialized marketing materials, create a new division and to hire outside experts. Less experienced or committed firms were more likely to engage primarily in staff training and to work with existing clients on LEED-related projects.

Most effective methods for marketing sustainable design services. Our survey participants were asked to describe the most effective methods for marketing sustainable design services, choosing from six options, or describing other approaches in their own words.

The most effective marketing means reflects the desire of building design professionals to let successful projects be their preferred marketing approach, which also reinforces the effect of networking, speaking and writing articles.

Marketplace effect of sustainable design focus. Of our survey respondents, 76% said that they had been able to attract new clients or projects based on their expertise, versus only 36% of the 2004 BD&C participants. This result points out the importance of developing expertise,

Table 6-1. Response to the Emerging Market for Sustainable Design Services

	Our Survey, Pct. of Total	BD&C, 2004,Pct. Of Total
An effort to LEED-certify at least one project	59	48
Create new marketing materials	56	24
Rely on in-house experts	53	63
Hire outside experts	27	18
Create a new division/profit center	12	6

Table 6-2. Most Effective Means for Marketing Sustainable Design Services

Means for Marketing Services	Percentage of Total
Successful projects, with LEED certification goal	37
Networking or speaking	18
Direct selling to interested prospects	12
Successful projects, without LEED certification	9
Public relations	8
Writing articles	6

project experience and a recognizable "name" in the early stages of a new market. In addition, 65% of our survey respondents felt that this expertise had helped them to retain existing clients, and 79% reported that this expertise and reputation had definitely helped them differentiate their firm and capabilities in the marketplace.

Of our survey respondents, 83% reported having attempted to sell clients and/or those in their organization on the virtues of using LEED on a particular project, versus only 54% in the 2004 BD&C survey. This point out the important role that developing internal expertise plays in convincing building professionals to "stick their necks out" and become advocates.

In our survey, of those who did attempt to persuade clients to do a LEED project, more than half (57%) are working on a LEED-registered project versus only 19% of the BD&C survey group. Again, if one asks for an opportunity to do something new and is trusted by the client, one if far more likely to get that opportunity.

Barriers to incorporating sustainable design and LEED into projects. In responding to this question about perceived barriers, Table 6-3 shows that our survey respondents gave more weight to first cost increases, found LEED projects harder to justify and found that the market was not willing to pay a premium for sustainable design. This may reflect their strong advocacy to all of their clients or an increased sensitivity to being turned down in an area of expertise.

Our survey respondents, more experienced in green design and more aggressive in promoting it, still found it hard to justify to clients, meaning that they were unable to connect their own personal or professional interests with the policy and project goals of their clients, and they found that the market was very uncomfortable with new ideas/technologies that might be involved in sustainable design. This suggests in some ways that incorporating sustainability and integrated design into the basic practice of a firm ("if you hire us, you get the following green measures, no discussion, no argument" approach) might be more effective and also help firms to differentiate themselves in the marketplace.

What can be done to more effectively promote sustainable design? In responding to this question about more effective marketing, Table 6-4 shows that our survey respondents gave more weight to independent cost information and less weight to case studies and more training than the BD&C respondents. Our survey respondents, perhaps more confident of their own abilities to sell projects, wanted to see more project experience

Table 6-3. Perceived Barriers to Incorporating Sustainable Design and LEED

Barriers	Our Survey, Pct. of Total	BD&C 2004 Survey Results, Pct.
Adds significant costs	78	52
Hard to justify, costs or otherwise	47	36
Not comfortable with new ideas/technologies	39	26
Market not interested; not willing to pay a premium	24	43
Other reasons	30	21

Table 6-4. More Effective Promotion, to Effectively Increase Comfort Level of Clients

Promotional Tools	Our survey, Pct. of Total	BD&C Results, Pct. Of Total
Case studies of successful projects	43	55
More independent cost information, in conventional formats (such as R.S. Means)	61	62
More training	28	60
More project experience	48	N/A
More successful local projects (ours or not)	43	N/A
Greater Life Cycle Analysis (LCA) of products	N/A	60
Better marketing materials	N/A	53

and more successful local projects. BD&C respondents wanted better information for product selection and better overall green building marketing materials. Both sets of survey respondents echoed need for performance and post-occupancy evaluations, usable and acceptable LCA tools, and a reduction in costs of certification.

Unfortunately, most design professionals are against selling their services per se. A number of survey respondents indicated that they would NEVER sell professional services—their idea of selling is to do a good job and hope someone notices. *They are not very good at selling*, either, in my experience, so that *this lack of presentation and persuasion skills presents a major barrier to more widespread adoption* of sustainable design. There is of course a major sales cadre of vendors who somewhat make up for this gap, by selling specific hardware solutions, but they seldom influence the decision for or against general green building approaches.

COMPETITIVE STRATEGY

In considering how to respond to new marketplace opportunities, professional marketers and businesspeople should take advantage of the past 30 years of strategic thinking by our large corporations and top business schools. Here, we present a brief introduction to some of that thinking, to guide a firm's strategic approach to this new market for green buildings.

GAINING COMPETITIVE ADVANTAGE

Most of this country's business schools teach some variant of the theory of business competitive advantage first introduced by Michael Porter of Harvard Business School about 25 years ago. Porter's classic work, Competitive Advantage (1980) first laid out the three basic building blocks of competitive strategy used by most businesses today.

In his work, Porter basically outlines three approaches to winning in the marketplace:

- Differentiation
- Low cost
- Focus

Differentiation

In differentiating professional services, one seeks to create a difference in the mind of the buyer, with attributes that make a difference. For example, we might want to be thought of as the "leading edge" firm; that will limit our market, but sharply define us to buyers who value that attribute, namely the "innovators" of diffusion theory. In today's commercial world, a major task for service firms is to create a *brand* that will incorporate and represent those key differences to potential clients. Of course, one can create differences for each market segment that one chooses to address, since some segments might value innovation, others low cost, others specific technological choices.

Low Cost

Low cost of operations gives a firm pricing flexibility, in that it can lower prices to "meet the market" and still make a decent profit. Given the tight budgets of many building projects in the U.S., the ability to compete on price is a valuable asset. In projects with a design-build delivery system, for example, lower costs to achieve specific defined sustainability goals may provide a winning edge for a construction firm. These costs may be based on prior project experience, good research or a willingness to "pay to get the experience."

An architectural firm that is really good at managing the process of integrated design might be able to design a LEED Gold-certifiable building at the same design fee and capital cost with which a less capable firm could only design a basic LEED-Certified building. The ability to be creative with green building "value engineering" for energy and water savings, along with high levels of indoor air quality, might help an engineering firm to create far more valuable green buildings for the same fee as a more conventionally oriented firm.

Low cost advantages might be more sustainable than even branding as a way to compete in the marketplace, but most firms don't have the discipline to operate in this fashion. As a good example of the competitive advantage of lower cost of operations, one can examine the almost unblemished success record of Southwest Airlines in making a profit while more prestigious companies enter bankruptcy. For Southwest, the low prices made possible by lower operating costs have become their primary brand, along with "fun."

Focus

Focus is a key competitive strategy, knowing which markets to compete in and which to shun, knowing which clients a firm wants and which it doesn't. Focus can be combined with low cost or differentiation as a strategy. Points of focus can include:

- Regional vs. national firm (many smaller firms compete nationally by narrowing their focus to one target market, such as museums, libraries, zoos and the like).

- Client types, which can include smaller clients, psychographic profiles (such as early adopter) or those distinguished by strong cultures and values of sustainability.

- Building types (or "vertical markets") such as office buildings, public service facilities (police, fire, jails), secondary education, higher education, health care, labs, cultural centers, retail, hospitality or industrial.

- "Signature" green measures, such as photovoltaics, Living Machines® or green roofs, a firm commits to bring into play on each project.

- Project size can also be a focus, allowing smaller firms, for example, to "fly under the radar" of larger and more capable competitors.

There is no one competitive response to the growing green building market that is "right" for every firm—as much has to do with the strategic clarity, capability, capital and character of the firm. Nevertheless, a conscious choice among strategies is vastly preferable to having none.

RECOMMENDATIONS FOR MARKETERS AND PRACTITIONERS

The following recommendations for green building practitioners and those organizations marketing sustainable design, while not surprising, follow from both industry surveys and from the well established theory of innovation diffusion, described further in subsequent chapters.

The marketplace wants and needs:

- Case study data, with solid cost information, including initial cost increments.

- Comparative cost information within and across building types, as to the full costs of LEED certification, including documentation.

- Demonstrable information on the benefits of green buildings beyond well-documented operating cost savings from energy and water conservation.

- Personal stories, by both practitioners and building owners, about the costs and barriers to completing LEED certified projects.

- Creation of a cadre of USGBC-certified building assessors who can provide certainty about the certification process, i.e., "if you do these things, your project will be certified", and a definable cost that can be included in project budgets from the outset.

- Stronger use of multi-media approaches and other modern sales tools, to increase the connection with green building goals and methods by stakeholders and decision-makers.

Practitioners need to understand how their marketing must evolve in order to compete effectively:

- They must pick a strategy that incorporates high levels of differentiation or low cost, with explicit focus on particular market segments, that might include geographic, project type, owner type, psychographic profile (e.g., early adopter, early majority), project size, a specific technological approach or "signature" green measures.

- This strategy must be reinforced internally and externally so that it becomes recognizable as a "brand identity" of the firm. Internal reinforcement includes training and certification of employees as LEED Accredited Professionals, for example; external reinforcement includes promotional activities to increase the visibility of the firm and its key professionals.

- Larger companies should consider developing their own proprietary tools for measuring sustainability, as part of a branding approach.

Along with these tools, firms should develop methods to successfully execute LEED projects without additional design fees.

- Architects and engineers must form closer working alliances with contractors and other project professionals to ensure that their designs will actually get built within prevailing budget, time, technology options and resource constraints.

THE TURNER SURVEY

The 2004 Turner Construction Company survey of more than 700 building industry professionals cited in Chapter 2 reported "executives at firms involved with more green buildings were far more likely to report that ongoing costs of green buildings were much lower than those not involved with green buildings." The main obstacles to widespread adoption of green buildings were found to be the following three, more or less in order of importance:

- Perceived higher construction costs (at a 14% to 20% premium!)
- Lack of awareness about the benefits of green buildings
- Short-term budget horizons for building owners and developers

Looking at these issues from a marketer's perspective, we can say that green building marketers are trying to sell a "product" or an "approach" that:

- Costs more
- Does not demonstrate significant benefits to balance the costs
- Must be sold to people heavily concerned about initial cost increases.

This is really hard work, as anyone experienced at all in sales and marketing can tell you! The solutions then become four-fold:

- Work aggressively to lower the costs of building green, through project experience and a focus on integrated design approaches that lower some costs (such as HVAC systems) while increasing others (such as building envelope insulation and better glazing) with a net positive cost impact.

- Rely heavily on case studies, testimonials from CEOs (and other believable business people) and make good use of the available academic research (such as the eBIDS system from Vivian Loftness and her group at Carnegie Mellon University, p. 59) that demonstrates the benefits of green buildings.

- Find ways to finance green building improvements to reduce or eliminate the "first-cost penalty" that often frightens away prospective buyers, using utility, state and federal incentives to maximize points of leverage.

- Rely heavily on studies for institutional buyers, fully 65% or more of the current market for LEED registered buildings, such as those of Kats, et al. (2003), that document the full range of green building benefits so that building owners with a long-term ownership perspective can be motivated to find the additional funds to build high-performance buildings.

The Turner survey showed that most executives and practitioners believe green buildings are healthier (86%), create higher building value (79%) and higher worker productivity (76%). They were more skeptical about such issues as higher return on investment (only 63% believed that), attracting higher rents (62% believed that to be the case for green buildings) and higher occupancy rates (only 52%), while only 40% of the respondents believed that green buildings in retail could bring about higher sales. The results of the survey are skewed even more when the relative experience of the respondents with green buildings is factored in; for example, 75% of experienced green building professionals believed that these buildings created a greater return on investment, vs. only 47% of professionals not experienced with green buildings.

One can draw the conclusion that the more marketing and production experience one has with green buildings, the more one is able to build a case, first in one's own mind, then in a client's mind, that this is the right way to go, and then to have the skills to execute one's intention to create high-performance buildings. At this point, the innovators and early adopters among the clients are ready for this strong advocacy—they are inherently more "sold" on the benefits of green buildings, less skeptical about their ability to achieve the desired results and more willing to work with design and construction teams to solve the problems that usually arise in trying new technologies and new approaches to building design.

CASE STUDY: TURNER CONSTRUCTION COMPANY

Turner Construction is the largest commercial construction firm in the United States, with annual revenues of more than $6 billion in 2002 (www.bdcmag.com, Annual "Top 300" survey), and annual revenues of more than $7 billion in 2004 (www.turnerconstruction.com). Since 1995, Turner Construction has completed, or currently has under contract, more than 85 green design projects with a construction value of $7.6 billion, as of 2005.

In late September 2004, Turner's CEO Tom Leppert announced a formal commitment to sustainable construction and business practices, as a means to continue strengthening Turner's leadership position. Leppert asserted that Turner's plan to be "the" (leading) responsible builder is good for the environment, and also for building owners, developers and occupants. Equally important, he stated that these practices are good for the bottom line and serve as an example to the entire construction industry. As the largest firm in the industry, Turner has effectively thrown down the gauntlet for other major construction firms wanting to compete with it. This as an extremely important development for the growth of the green building industry, since most sophisticated building owners and developers rely heavily on the advice of their builders in deciding to adopt green building design for their projects. Leppert stated:

"As our experience in green building has grown we've learned that costs, contrary to common belief, can be contained to a level comparable to traditional, non-sustainable buildings and generate additional, important benefits for our clients and our local communities. Turner plans to leverage this experience and increase its already-broad involvement with green practices for the advantage of our employees, our clients and the environment."

Turner's Commitment to Green

The proposed *Turner Green* program consists of:

- Mandating construction site recycling on all Turner projects, not just green design projects. Recycling efforts will be phased in until Turner reaches 100% of its projects.

- Ensuring that over time, all Turner field offices will be green-friendly. In these buildings, Turner will incorporate field waste-recycling programs, energy-efficient lighting on timers, operable windows for natural ventilation and water-efficient fixtures.

- Implementing a collaborative sponsorship with the USGBC of the "Emerging Green Builders" program, to help improve sustainable building curriculums at colleges and to recognize those students who will continue to promote future of green building growth. (See Chapter 19 for a discussion of the importance of recruiting future employees with a green commitment).

- Instituting a major green training program for Turner employees. Turner's online tool, *Turner Knowledge Network*, will help employees learn about the LEED standard, to add to their knowledge of green field operations guidelines. (In our view, this internal training role is critical to the marketing of the green capability and is often overlooked, especially in the construction field. Without internal training, it is difficult if not impossible for a company to "walk the talk," as discussed at greater length in Chapter 10, as "Internal Marketing".)

- Doubling the number of Turner's LEED APs from 42 to 84 by the end of 2005. (While this is not a large number for such a large firm, it is a beginning.)[28]

- Creating an advisory council of outside industry experts to provide objective advice on sustainable design best practices and to drive their adoption with the company and its clients.

- Naming a Senior Vice President (Rod Wille) to lead Turner's *Center of Excellence*, to link Turner's local and national green information. Wille has been in this role informally for several years, and his inside knowledge of the company and credibility within the company have materially advanced prior green building initiatives.

"From now on, whenever businesses consider undertaking a new building project they should first think green, and then think of Turner because we have the resources, the experience and the knowledge to do green right," Leppert said.

Turner's Green Experience

One of the projects Turner completed in 2004, The Genzyme Center (see Chapter 11) received a LEED-Platinum rating in 2005. Within Genzyme's budget, Turner was able to incorporate innovative features including sun-tracking mirrors to direct daylight into the building, natural ventilation using the atrium, and a double-skin exterior wall and extensive indoor gardens for the enjoyment of occupants and to improve indoor air quality. During procurement, Turner helped Genzyme and the design team ensure that the contract documents incorporated the green elements desired by Genzyme and that subcontract bidders used cost-effective products and methods to achieve the LEED Platinum rating within the budget constraints.

Also in 2003, Turner was able to partner with Toyota to develop a LEED Gold-Certified building in Torrance, Calif., that cost no more than a traditionally constructed building. The *Toyota Motor Sales—South Campus* building, completed last year, is 636,000 square feet on a 38-acre site. For use as administrative offices, it features 53,000 square feet of rooftop photovoltaic panels that can generate 550 kilowatts of electricity—or about 20 percent of its total energy usage. Its first cost was competitive with the cost of other local, conventional office buildings.

"The expected increase in green building benefits us all, especially Turner clients," said Leppert. "It streamlines processes and controls up-front costs for construction while ensuring that sustainable methods will be used whenever possible."

Chapter 7

Vertical Markets for Green Buildings

In this chapter, we address several selected "vertical" markets for green buildings, i.e., markets that are already developed or that are expected to develop rapidly. These include commercial offices, K12 education, higher education, public facilities, housing and healthcare (a still developing market). In this terminology, a "vertical" market refers to housing a particular type of use for a building—office, education, medical, etc., whereas a "horizontal market" refers to green technologies that could be used in a wide variety of building types, for example, solar energy systems can be used in offices, schools, churches, etc. Vertical markets for green buildings exist in every area of the country, so it makes sense to look at how these markets view green buildings at the present time and how marketers are trying to address the needs of particular building types.

It pays to remember two key facts when addressing each of these markets: relatively few architects have designed a LEED-certified building (as of the end of 2005), and few owners have purchased one. Therefore, we are still very much in the "early adopter" stage of market development. A few building owners are now putting out requests for proposals that do specify that a building project must achieve a LEED rating, but only a few selected government agencies are demanding a LEED Silver or Gold rating. Some nonprofits are even going so far now as to specify that a new project has to achieve a Platinum rating, but these are still very rare occurrences.

COMMERCIAL/OFFICES

Commercial and office construction represented a $115 billion market (annualized) in 2005, with offices directly accounting for $45 billion. According to USGBC data by building type (September 2005), 15% of LEED-registered projects were commercial offices and 31% were "multi-use," a category that includes commercial offices with, for example, ground-floor retail, parking garages or other uses. (The multi-use category may include, for example, housing with retail, and other forms of multiple use that do not include commercial offices). Given that LEED is most clearly usable as a green building design and rating tool for office buildings, it's no surprise that commercial offices would comprise over 25% of the total registered projects. These office buildings include projects for private companies, major corporations, developers, government agencies and nonprofit organizations.

Of the first 246 LEED-certified projects listed on the USGBC web site as of October 2005, 78 (32%) appear to be some form of office project. So, clearly the market for LEED projects is still highly concentrated in the easiest market to approach, which is office buildings. It is easy and fairly inexpensive to certify an office building project under LEED guidelines, with most projects we have seen getting 20 or more LEED points just in their initial basic design. Certification costs might run $60,000 to $75,000 for a typical 100,000 sf building, including documentation, energy modeling and building commissioning, or about $0.60 to $0.75 per sq.ft., at the basic level of LEED certified.

A good example might be the IBM Tivoli Systems headquarters office in Austin, Texas, completed in January, 2002, and receiving 26 points, the bare minimum needed to certify under LEED. The project is a 5-story urban office building with 200,000 sq.ft. of space, serving about 750 people. According to the project case study, "the green strategies added about 3% to the total cost of the building's core construction cost, compared with IBM's normal building standards." Given that this building was designed in 2000, before architects and engineers were really familiar with the LEED system and before all the "bugs" in the system had been worked out, we believe *a similar building today would show little or no construction cost premium* to meet the basic LEED certification requirements. For a project of this size, one would expect a cost of about $0.50 to $1.00 per sq.ft. for meeting all certification requirements, including energy modeling and building commissioning, as well as certification documentation.

This conclusion (of no major construction cost increases for a LEED certified office building) is supported by a report issued in 2004 by the international cost consulting firm, Davis Langdon, "Costing Green: A Comprehensive Cost Database and Budgeting Methodology[29]". Based on a study of hundreds of LEED and non-LEED buildings, the study concluded that construction "cost differences between buildings are due primarily to program type," and not to the green features included in the building (report at page 23). This conclusion does not mean that more "exotic" or "add-on" green building measures (such as photovoltaics or green roofs) would not add significantly to the capital cost of a project, or that there are not additional costs associated with LEED certification of a project, but rather that it is not possible to ascribe additional capital costs with certainty to LEED projects, since other program factors are far more dominant contributors to building costs.

Another 2003 study by Greg Kats, referenced earlier (page 40), reported to over 40 California state agencies and provided the first rigorous assessment of the costs and benefits of green buildings[30]. Drawing on cost data from 33 green building projects and financial benefits data from over 100 buildings nationwide, this report concluded that LEED certifications add an average of 1.84% to the construction cost of a project, and that one could expect to pay about 6.5 % more for a LEED Platinum project. Given that there are less than 10 documented LEED version 2.1 "Platinum"-certified projects in the U.S., as of September 2005, it is impossible to tell what premium a Platinum project would carry, but most sources assume it would be in the range of 5% to 10%. For Gold-certified office projects, most observers expect a construction cost premium in the range of 2% to 5% over the cost of a "code" building at the same site. Cost premiums will also vary for new construction vs. renovation, and urban vs. suburban locations, among many other factors, including variations in local and state code requirements. (Extra costs may be justified as increasing a building's value by reducing its operating costs, for example).

About 37%, 17 out of 46, of the first commercial office projects certified under LEED, were either built by or for public agencies, slightly below the 44% share of all registered projects belonging to local, state or federal government agencies. Adding in public safety facilities and most of the cultural and recreational projects (15 in total) would bring the share of the initial 102 certified projects belonging to public agencies a bit closer to the percentage of registered projects belonging to this owner type. (One explanation for the discrepancy between publicly-owned registered and certified projects might also lie in the treatment of higher education projects by the USGBC database.) Including the nine publicly-owned schools and universities would bring the number of publicly-owned or used certified projects up to 40% of the total. Another explanation is that public agencies have really stepped up their commitment to LEED in the past two years and that many of their projects are still in construction or in the midst of the certification process, so that an examination of the database in 2005 could yield the result that publicly-owned or occupied LEED certified projects are nearly 45% of the total.

So if one is a developer or builder of "green" commercial offices, it would pay to be aware of and connected to the public agency market, which will likely represent nearly half of all commercial offices to be built to LEED standards in the next few years. Another good reason for staying on top of this market is that many public agencies are adopting "LEED-friendly" policies for their new commercial building projects.

EDUCATION

The value of educational construction exceeded $75 billion (annualized) in 2005. Imagine that this market consists of 5,000 to 10,000 buildings valued at $7.5 to $15 million each; now further imagine that eventually 1,250 to 2,500 of those will be LEED registered each year, given that LEED aims to address the top 25% of the market in each building sector. Therefore, in 2005, we can predict that about 14% of all LEED-registered projects will be from the education market segment, or about 85 projects (14% of 600 newly registered projects), representing perhaps a 4% penetration of the ultimately accessible market for LEED education projects. Using our terminology from diffusion theory, we are now clearly in the "early adopter" phase of this market, and we saw signs of accelerating growth in 2005.

K12 Education

Of LEED-registered projects as of September 2005, only 6% were specifically dedicated to the primary or secondary schools. This is surprising in light of the overwhelming evidence favoring daylighting and views of the outdoors, for example, in schools (to improve student health and raise test scores, see studies cited at www.h-m-g.com) and the need for schools to keep long-term operating costs for energy and water under control. A proprietary survey conducted for the author in 2001 found, however, little interest or awareness of sustainable design or green buildings among West Coast school administrators. Based on nearly 100 detailed phone interviews, that survey concluded that most school people (principals, school board members, facilities directors and business managers) were expecting their design teams to bring these approaches to them, and that most design teams were expecting the schools people to suggest them! (This is not surprising, since school board members are almost always volunteers, mostly without special expertise in building design or construction.) Our experience suggests that the school funding process is also a significant factor: most schools are funded by general obligation bonds, with a process that can take one or two years to develop costs and funding proposals, with another year or more before voter approval. In this situation, most schools being built today were conceived before LEED version 2.0 was developed, and so did not have a chance to incorporate those costs and measures in the cost estimates or justifications for the schools. Also, green design has not been a high priority for those architects most closely involved with school design (who tend to be smaller local firms) until very recently.

Nevertheless, there are exemplary schools that are being built to LEED standards, typically by visionary architects and school district superintendents. For example, in the Portland, Oregon, metropolitan area, the LEED Silver Clackamas High School, was completed in the spring of 2002, at a cost of $118 per sq.ft., below the average cost of other local high schools at that time. A second project by the author's engineering firm, the Corvallis, Oregon, high school, was completed in 2005, also on a conventional budget, and is currently seeking LEED Silver certification.

Other LEED-certified schools projects are in New Mexico, North Carolina, Pennsylvania, Michigan and Virginia. All told, there are about 133 additional K12 schools registered under LEED and in various stages of design, construction and certification, as of the end of September, 2005, based on 6% of nearly 2,200 LEED registered projects.

School design tends to be a rather specialized field, and one must depend on architects who already design a lot of schools to lead the way in "greening" school construction. Some of these architects are leaders in green design, but our experience is that most are still "feeling their way" into this new area of design and construction. Most school districts are still trying to understand the budget and schedule implications of setting LEED goals for their schools. In areas of the South, Southwest and West Coast with rapid student population growth, there is considerable pressure just to build "anything" that will be ready for a September opening in two years and that will fit within a budget that might have been "sold" to the school board and the community several years prior. In fact, until recently it has been rare, in the author's experience, to see a school district in Oregon or Washington (two prime areas for LEED-registered projects) issue a request for qualifications for architects that includes sustainable design experience or expertise among its scoring criteria. As the saying goes, "what gets measured, gets managed," and one might add, "what gets evaluated, gets selected." However, there are signs of change in 2005, and we are beginning to see some requests for qualifications (RFQs) awarding 5% to 10% of total evaluation points for design teams with LEED project experience.

It is very likely that school design will begin to include more and more green design measures, such as daylighting, low-VOC materials, higher levels of energy conservation and water conservation, and recycled-content materials, before we begin to see a sharp increase in the number of LEED-certified or even LEED-registered school projects. So marketers and design professionals should be spending their time trying to sell the benefits of green design to school boards, administrators and school facilities people, while remaining aware that LEED-registration and certification for new projects may be a long time coming in this market sector. In addition, there are competing standards, most notably the CHPS program of the Collaborative for High Performance Schools (www.chps.net) that is increasingly influential in California and (as of 2004) in Washington state. On its web site, CHPS publishes free "Best Practices" manuals that represent a major advance in the green design of schools and are essential references for architects, engineers and builders of schools.

Higher Education

According to the LEED statistics, higher education projects comprise 7%, or slightly more than 150 of the first 2,200 LEED-registered projects, through the end of September, 2005. Since college housing is now a very large and growing market, with the explosion of college registrations since 2000 expected to last through 2009, one can expect a significant number of LEED projects in higher education will involve student housing. With Mithūn architects of Seattle, the author's firm helped design one such project at Portland State University, a five-story dorm, Epler Hall, now Silver-certified. Another project in construction at Humboldt State University in Arcata, California, is a contractor-led project that aims to achieve a Gold level.

The main market for higher education projects involves such facilities as:

- Classrooms
- Offices
- Libraries

- Performing arts
- Laboratories and research buildings
- Student housing (often with some classroom space)
- Recreation centers and college athletic facilities
- Student centers, and combinations of these facilities.

Assuming there are about 3,000 colleges in the United States, and that each institution starts an average of one building project per year, LEED-registered projects are currently less than five percent of the college/university market. As of September 2005, there were only 44 certified higher education projects in the entire country, so market penetration in this sector is just beginning. In the campus environment, at least 50 percent of the LEED projects exist due to strong support from the institution's top leadership.[31]

The author's firm was part of a successful contractor-led team for a California State University system project that had about 15% of the total project evaluation based on energy conservation and green design measures. The team "guaranteed" a LEED Gold project as part of its winning proposal, including daylighting, natural ventilation, high levels of energy conservation, 100% rainwater capture and recycling, and considerable use of recycled-content and low-VOC building materials. This project began construction in the summer of 2004 and is expected to be ready for occupancy in 2006. It is expected to be the first LEED Gold-certified building in the 23-campus California State University system. (The 10-campus University of California system and the dozens of local community college districts in California run separate systems and have begun to include more and more green building requirements in their design and construction programs).

The role of various stakeholders makes the college and university market markedly different from the K12 education market. In higher education, the students and faculty are far more influential, with sustainability being a major "buzzword" on campus these days. As a result, green buildings in higher education are starting to acquire momentum as a force in the design of new structures. These buildings also offer many opportunities to incorporate green buildings into the curriculum, involving multiple departments such as environmental studies, architecture and engineering. We are seeing considerable faculty interest in 2005 at hundreds of colleges and universities in getting sustainability issues and considerations into coursework and research.[32] Some university administrators are also beginning to see opportunities for green building programs to assist with fundraising and with student and faculty recruitment. See, for example, "Campus Sustainability and Green Building" at the web site for the Environmental Studies Department at 1,800-student Lewis & Clark College in Portland, Oregon[33].

Higher Education Green Building Survey

In the spring of 2004, the author conducted a survey of more than 1,000 college and university administrators, faculty and facilities directors, with more than 150 of them responding to a web-based survey, 75% of them representing public universities, and 42% representing schools with more than 20,000 students. The database was primarily drawn from two professional organizations representing campus planners, architects and facilities directors. Half of the respondents were campus architects or facilities managers, while only 13% were faculty. Of the total, 47% of the organizations were members of the U.S. Green Building Council, and, of those non-USGBC members, only 10% had plans to join in the near future. So, in general, the respondents represent a typical buyer for green building services, i.e., a large public university with a variety of project goals, not just LEED certification. Table 7-1 shows the various types of projects this respondent group has been building since 2001.

Table 7-1. Higher Education Project Types Built or in Design (2001-2005)

Type of University/College Project	Involved with this project type Percentage of Respondents
Classroom/office	70
Labs/research	52
Housing (low-rise)	42
Recreation/student center	29
Libraries/museums	25
Housing (high-rise)	19
Administration buildings	15
Other (including parking garages)	23

When asked whether projects had sustainability goals, 89% of the respondents said "yes." The goals ranged from green goals in the building program, to green purchasing policies, specific LEED goals, and tie-ins to specific campus programs such as composting and recycling. Energy conservation and recycling were key factors in nearly 90% of the projects. Half the respondents had campuses with coursework in sustainability, and nearly half had specific LEED goals, formal mission statements about sustainability, and some type of sustainability committee.

From a marketing point of view, 80% of the survey respondents identified the facilities director and department (along with a campus architect who is frequently situated in that area) as instrumental in these programs and goals, with 60% identifying top-level administrators, 59% students and 54% faculty. This survey clearly shows the role of key stakeholders from the faculty, students and staff in influencing decisions to "go green" at the campus level. Interestingly, 50% of respondents said that top-level support was strong or fairly strong for their green building programs. Top level support was strongest at the smaller public and private institutions, where one might expect the chancellor, president or provost to be more actively involved in all aspects of campus life.

Of those respondents (93%) with active building programs underway, Table 7-2 shows the specific sustainability goals identified.

Table 7-2. Specific Sustainability Goals in Higher Education Building Programs

Sustainability Goal	Percentage of Projects with this Goal
Daylighting	64
LEED Certification	62
Use recycled-content materials	53
Greater than 30% energy savings	48
Greater than 30% water savings	35
Recycle construction waste	36
Renewable energy (photovoltaics)	17

Energy issues, as seen in daylighting, energy conservation goals and use of renewable energy are quite important in these projects, as is involvement with recycling, both in terms of construction and demolition debris and in using recycled-content materials. LEED certification is a goal for a majority of the projects mentioned. In terms of design process, 52% reported conducting a design charrette or sustainability forum as part of a green building project.

When asked about barriers to implementation, respondents to this survey voiced similar concerns to those in the 2003 survey cited in Chapter 6. Table 7-3 below shows what this group of buyers and owners cited as barriers to implementation of green design goals, practices and technologies in their building projects.

Table 7-3. Barriers to Implementation of Sustainability Goals in Campus Projects

Barrier Identified	Percentage of Respondents Citing It
Increased costs, real or perceived	87
Not seen as an administration priority	31
No prior experience with green design	23
No strong campus constituency	18

In other words, *the overwhelming barrier to implementation* is real or perceived cost increases. Often, the facilities group on a campus is given fixed budgets, to which they may not have had sufficient input. Therefore, increased costs are a key project construction barrier for them in delivering the project with the program demanded by the administration and faculty. Other barriers cited included high "soft" costs for LEED documentation and required services, local building codes, project schedules and other time constraints; difficulty of mixing capital and operating budgets to justify balancing the higher initial cost of energy conservation investments with future savings, and poor timing of introducing green goals or sustainability values into a project.

When asked how to increase their comfort level with green building goals, processes and technologies, 61% of the respondents wanted cost information in standard formats such as RSMeans[34], while 58% wanted such standardized cost information on specific green building elements, such as green roofs, photovoltaics, energy efficiency measures and the like. Nearly half (46%) cited the need for more of their own experience to feel comfortable, while more than 40% wanted to see detailed case studies of university projects and/or local green building projects they could learn from. More than a third wanted specific information on the cost of LEED projects, particularly at various levels of certification.

As a final guide to marketers, the survey respondents were asked to comment on how they would approach sustainability in future projects. Several suggested that they were going to be adding sustainability to campus planning as a guiding principle in the near future, and that they would add sustainable design criteria into the overall design guidance for future projects. (In fact, we are seeing increasing evidence that there are active Sustainability Task Forces now at many major universities). The main difficulty cited in their comments for investments in energy efficiency, for example, was the separation between capital and operating budgets and the difficulty of getting additional capital appropriations for improvements that go much beyond code.

PUBLIC FACILITIES (OTHER THAN OFFICES)

The market for green buildings for public agencies is perhaps the largest single green market in the U.S. and is growing rapidly. (Taking a third of the office market, public safety and recreation segments alone, this market exceeds $43 billion per year, much of it in smaller buildings.) Whether for office buildings, public safety and order, cultural or recreational projects, or even public housing, there is a growing green building market driven by public agencies, including local, state and federal, to meet an ever growing array of public policy pronouncements in favor of achieving LEED certification for all new building projects, typically those above 5,000 sq.ft. The author recently advised a design team, for example, on two LEED-registered projects, a new police headquarters and a renovated vehicle maintenance facility for a major Northwest city, which are in construction and on track to receive LEED Gold and Silver certifications, respectively.

Types of public agency projects with LEED goals often include:

- police stations and emergency communications centers (9-1-1 facilities)
- fire stations (in 2004 the City of Issaquah, Washington, certified a new fire station at LEED Silver)
- forensic labs
- pools and recreation centers
- community centers and senior centers
- museums and libraries, visitor centers
- performing arts centers
- city halls and county centers
- convention centers
- administrative buildings of all kinds
- airports
- courtrooms and jails
- warehouses and vehicle maintenance facilities
- public housing.

In terms of building size, the largest projects tend to be those for the federal government, followed by state government buildings. The U.S. General Services Administration has been one of the leaders in adopting LEED and pushing it into their projects through a "Design Excellence" program. The federal government budgeting process also seems conducive to using green building measures, since the "feds" have the attitude of a long-term owner-operator of buildings and a long-standing commitment to energy conservation in buildings via the Federal Energy Management Program (www.eere.energy.gov/femp).

HEALTHCARE AND HOSPITAL FACILITIES

This is a potentially large market that is still in its early stages of development. Currently less than 3% of the LEED-registered projects represent medical or healthcare facilities. The first (and to date, only) LEED-certified healthcare project, Boulder, Colorado, Community Foothills Hospital came on line in 2003, rated at LEED Silver. Examining the point totals showed no water conservation points, only 30% energy savings vs. ASHRAE standards, but considerable attention to attaining Indoor Environmental Quality and Materials credits. Currently available are the Green Guidelines for Healthcare Construction (GGHC), available at www.gghc.org. Version 2.0 of these LEED-related guidelines was issued in the Fall of 2004. The guidelines cover both construction and on-going operations, similar to the LEED-NC and LEED-EB standards. A pilot program for evaluating projects will continue through 2005. It is unclear at this point how the LEED rating system and GGHC rating system will eventually relate to each other. [35, 36] But if one were to hazard a guess, hospitals and healthcare architects are likely to go their own way with the GGHC , although there are still strong forces holding the two together.

However, what is clear is that architects and facility owners (85% of health care facilities are owned by nonprofits) have a strong stake in creating healthier environments for doctors, staff and patients. Some larger owners, such as the Kaiser Permanente (KP) in California, have already aggressively begun to address green building and green operations issues.[37] Describing Kaiser's commitment, one article states:

The California Sustainable Hospitals Forum, convened in June 2003, assembled architects, designers, engineers, owners and contactors to discuss how best to incorporate environmentally sustainable practices into healthcare facilities. The timing of the forum coincided with the tremendous hospital building boom in California, driven by seismic safety codes that require adherence to more stringent standards. KP and its partners seized this opportunity to study and proliferate ecologically superior building designs throughout the industry.

Therefore, this market bears watching; if your firm is already active in the healthcare market, you need to start paying attention to these guidelines and making them part of your approach to hospitals, clinics and medical offices. According to the data presented in Chapter 4, healthcare is a $37 billion annual construction market (2004), more than four times the religious or public safety markets, almost one-third the size of the office building market, and about 50% the size of the education market. So it is a large and ever growing segment of the market. Construction spending on healthcare facilities is expected to increase 10% in 2005 (Building Design & Construction, September 2004, p.11, www.bdgmag.com). In California, for example, hospitals are "under the gun" to complete most seismic updates by 2008 and 2013, and it appears that this renovation and upgrade market would be an ideal one for green buildings, in terms of retrofitting energy use, water use and indoor air quality systems, as well as adding daylighting, recycled content and less toxic materials.

HOUSING

As a vertical market for green buildings, housing is just starting to develop. Multi-unit (above three stories) residential LEED registrations are running at about 3% of the total, or just under 70 of the initial 2,200 registrations through September 2004. The first LEED-Gold high-rise (apartment) residential project, *The Solaire,* in New York City, was certified at the end of 2003. Based on September 2005 USGBC web-site data, LEED has currently certified only four other private sector housing projects, both at the basic (Certified) level, and about eight student residences at campuses, so this segment of the market is still early in the development or "innovator" stage. (The Epler Hall project, LEED Silver, at Portland State University is also a dormitory.)

Chapter 12 addresses specifically single-family developments and multi-family markets for both rental and ownership units, and Chapter 16 addresses the future of green building rating and certification systems for the low-rise and single-family detached housing market. A developing green building market is for student housing, particularly developer-led projects. There are a number of nonprofit (and for-profit) organizations in this marketplace. For example, Cal Poly State University in central California in 2004 issued an RFP for a developer to build 2,600 units of student housing. Other projects we know of (just in Oregon) are at Portland State University, Oregon State University, Eastern Oregon University, and the Oregon Institute of Technology.

The project shown here, Epler Hall at Portland State University, completed in 2003, is a LEED Silver project with 123 residential units on five floors. The project is 35% more energy efficient vs. local code and recycles 26% of its rainwater for flushing toilets in the first floor public use area, and provides extensive daylighting. It received its LEED certification in October of 2004. Projects such as Epler Hall are becoming increasingly common on university and college campuses. They offer a way to attract students and to promote the school's commitment to sustainability.

Epler Hall, Portland State University, Portland, Oregon, (*Courtesy of Interface Engineering, Inc.*)

Chapter 8

Demand for Green Building Measures

PART ONE

DEMAND FOR SPECIFIC GREEN BUILDING MEASURES IN LEED-CERTIFIED PROJECTS

Using statistics from the U.S. Green Building Council, we can profile specific green building measures that are used by the green building market. For the first 195 LEED certified projects, USGBC data indicate specific measures used, as shown in Table 8-1. Note that the current split of LEED version 2.0 certified projects is 45% certified (88 out of the first 195 certified projects), 31% Silver, 22% Gold and 3% Platinum. Higher levels of certification demand more of specific green building measures.

Use of Green Building Measures in LEED-certified Projects

Tables 8-1a and 8-1b help understand not only how to achieve LEED points, but which measures are likely to be used in green building projects. The use of specific green building products and green design measures generally falls into three distinct categories (Table 8-1a). As the market for higher levels of LEED certification grows, we can expect that certain products in the "somewhat likely" category will be used in more than 67% of projects, such as CO_2 monitors, and that certain products such as photovoltaics (even though the cost/benefit ratio is high) and FSC-certified wood will move into the "somewhat likely" category, because they are more "visible" signs of commitments to sustainable building measures than others.

Based on the data in Table 8-1b, we estimate in Table 8-2 the market size for various green building measures for a "typical" year in which 1000 projects register for LEED certification. This may occur as early as 2008, according to Tables 4-5 and 18-1. Part Two of this Chapter deals with the process of creating a high-performance building. These buildings achieve high levels of energy efficiency without sacrificing indoor air quality or thermal comfort, in most cases. Following the section on energy-efficient building design, this chapter deals with commercial and institutional buildings employing solar power systems, using both building-integrated and stand-alone photovoltaics systems.

This book does not deal specifically with marketing green products in commercial buildings, products that assist in meeting exacting requirements for points in several LEED credit categories, including water efficiency, green roofs, low or no-VOC materials, high-recycled-content materials, Energy Star roofs, certified wood products, and materials made from rapidly renewable materials such as cork, bamboo and agrifiber products. Many of the other measures that receive LEED points, as listed in Table 8-1, involve design and construction decisions that are made at various stages of the integrated design and building process, including specifications and other construction documents, and do not require specific marketing measures by outside firms. They are more likely to be influenced by the project's LEED goals, by the use of an integrated design process, and by the relative green design skills of the firms involved.

So, even in this brief rendition, we can see that identifiable green building measures in LEED registered buildings may account for nearly $800 million in new market value in 2006. When we add in the large expenditures for energy efficiency measures with relatively quick paybacks, there may be billions of additional dollars spent on green materials and systems, much of it replacing expenditures on "less green" items, stemming from projects' decisions to increase their level of sustainability.

PART TWO

MARKETING ENERGY-EFFICIENT TECHNOLOGIES TO BUILDING OWNERS AND DEVELOPERS

Owners and developers of commercial buildings are discovering that it's often possible to build high-performance, energy-efficient buildings on conventional budgets. For the past ten years, and particularly in the past five years, in ever increasing numbers, we have begun to see development of commercial and institutional structures, using green building techniques and technologies.

Table 8-1a. Green Measures Used in LEED Projects

Highly likely to be used (67% or more of projects)
Low VOC paints, coatings, adhesives, sealants
Low VOC carpeting
10% or more recycled-content materials
Views to the outdoors from 90% of spaces
Two innovation credits: public education, high levels of construction waste recycling, etc.
Somewhat likely to be used (33% to 66% of projects)
Two-week building flushout prior to occupancy
Carbon-dioxide monitors to improve ventilation effectiveness
Bioswales, detention/retention ponds and rainwater reclamation systems
Green roofs or Energy Star roofs
Construction-period indoor air quality maintenance
Permanent temperature and humidity monitoring systems
Daylighting for at least 75% of spaces
Cutoff light fixtures and lower outdoor ambient lighting levels
Water-conserving fixtures and waterless urinals (30% reduction in 54% of projects)
At least a 35% energy use reduction over ASHRAE-modeled levels
Additional building commissioning: peer review of design-phase documents
Purchased green power for at least two years
No added urea-formaldehyde (UF) in composite wood or agrifiber products
Less often used (less than 33% of projects)
Solar photovoltaics (9% to 13%)
Electric vehicle charging stations/Alt. Fuel Vehicles (31%)
Measurement and verification systems, using U.S. Dept. of Energy protocols (30%)
Site restoration with natives (29%)
Use of FSC-certified wood products (26%)
High-efficiency ventilation, including underfloor air systems (25%)
Operable windows (30%)
Rapidly renewable materials, such as cork, bamboo, agrifiber boards, linoleum products (6%)

Understanding High-performance Buildings

Since its introduction, LEED has captured about 3% of the total new building market, with more than 2100 "registered" projects (as of September, 2005) encompassing more than 247 million gross sq. ft. of new and renovated space. Since a project only gets "certified" under the LEED system once it is completed and ready for occupancy, many projects are just coming up to the finish line of completing the documentation for

Table 8-1b. Specific LEED points used by LEED certified projects[38]

LEED Credit Category	% of Projects Certifying	Typical Measures Used to Meet Point Requirements
SS 4.3 – Alternative fuels	31%	Electric vehicle charging; Hybrids
SS 5.2 – Site restoration	58%	Preserve habitat; use native vegetation
SS 6.1 – Stormwater management	37%	Bioswales; detention ponds; rainwater capture and recycling
SS 7.2 – Urban heat island effect	46%	Green (vegetated) roofs; Energy Star roofs with high emissivity
SS 8 – Light pollution reduction	46%	Cutoff fixtures; lower ambient lighting
WE 3.2 – 30% Water use reduction	54%	Low-use fixtures; waterless urinals
EA 1.1 to 1.5 – Project achieved 35% reduction vs. ASHRAE 90.1-1999	42%	High-performance glazing; reduced ambient lighting levels; better envelope
EA 2.2 – 10% renewables for electricity use	9%	Photovoltaics; on-site renewables
EA 3 – Additional commissioning	47%	Third-party commissioning
EA 5 – Measurement/verification	30%	Additional energy monitoring
EA 6 – Purchased green power	36%	Buy green power for two years
MR 4.2 – 10% Recycled content materials	67%	Specify recycled-content materials
MR 6 – Rapidly renewable materials	6%	Cork, linoleum, agrifiber MDF board
MR 7 – 50% of wood products certified	26%	FSC-certified lumber
EQ 1 – Carbon dioxide monitors	55%	CO_2 monitors
EQ 2 – High-efficiency ventilation	25%	Underfloor air systems
EQ 3.1 – Construction indoor air quality	54%	Best practices/ MERV-13 filters
EQ 3.2 – Air quality prior to occupancy	56%	Two-week flush-out
EQ 4.1/4.2 – Low-VOC paints/adhesives	80%	Specify low-VOC materials
EQ 4.3 – Low emission carpeting	93%	Specify low-VOC carpeting
EQ 4.4 – No added UF in composite wood	41%	No added UF in composites
EQ 6.1 – Thermal comfort at perimeter	30%	Operable windows
EQ 6.2 – Thermal comfort interior	20%	Underfloor air systems
EQ 7.2 – Temperature/humidity monitoring	60%	Humidification/dehumidification
EQ 8.1 – Daylighting for 75% of spaces	39%	Light shelves; skylights
EQ 8.2 – Views to outdoors - 90% of spaces	68%	Space layout; larger windows
ID 1.2 – Two innovation credit points	79%	Too many to mention
ID 1.4 – Four innovation credit points	29%	Too many to mention

a LEED rating. In 2003 and 2004, three projects in southern California achieved the highest "Platinum" rating under LEED, the first such U.S. projects to be so certified; two of these projects were for public agencies: a Riverside, California utility and Los Angeles County, while the other was for a nonprofit. As of September 2005, more than 280 projects had completed the certification process under LEED, and the number of such projects is increasing at a rate of about 12 to 15 per month.

What are the differences between using other guidelines and using the LEED process? In one sense, they are complementary: using other energy and indoor air quality-specific guidelines can take a project more

Table 8-2. Estimated Minimum Annual Market for Green Building Measures in LEED-registered projects (actual market may be much larger for some)

Green Building Measures	Percent of Projects Using Measure[39]	Pct. of Total Materials Cost	Estimated Market Value in 2006 (1000 projects)
Recycled content	67%	10%[40]	$377 million
Rapidly renewable materials	6%	5%	$ 17 million
Certified Wood	26%	1%[41]	$ 29 million[42]
Low VOC paints, sealants, adhesives, etc.	80%	0.5%[43]	$ 22 million
Low VOC carpet	93%	NA	$150 million[44]
Photovoltaic systems	9%	N/A[45]	$ 72 million[46]
Green roofs	10%	N/A	$ 15 million[47]
Underfloor air systems	20%	N/A	$ 120 million[48]
Waterless urinals	40%[49]	N/A	$ 2 million

Table assumptions: 1000 LEED Registrations, average project size 100,000 sq. ft., project construction cost @ $125/sq. ft., materials cost at 45% of construction cost (default value in the LEED calculator), giving an estimate of total materials cost at $5.63 billion for LEED projects.

than halfway toward LEED certification. For example, if the focus is primarily on energy use, reducing carbon dioxide emissions (linked to global warming) and improving indoor air quality, then metrics such as the Advanced Building guidelines can take a project there efficiently. These improvements lead to reducing operating costs and improved occupant health, productivity and comfort. These guidelines also reduce design costs by giving clear performance benefits for specific design measures.

Both LEED and other building evaluation systems encourage an "integrated design process," in which the building engineers (mechanical, electrical, structural and lighting) are brought into the design process with the architectural, civil and structural team at an early stage, often during programming and conceptual design. Integrated design explores, for example, building orientation, massing and materials choices as critical issues in energy use and indoor air quality, and it attempts to influence these decisions before the basic architectural design is fully developed. The author developed a system of 365 questions to guide an integrated design process, organized by design phase, to ensure that good choices were not precluded by simply not being considered at the right time during the fast-track design process that characterizes most projects today[50].

Marketing High-performance Buildings

It is indeed possible to market smaller LEED building projects to owners and developers. There are many examples of small office projects that have achieved a LEED-certified rating on a conventional budget, ranging from an owner-occupied, 15,000 sq. ft., three-story office building in Lake Oswego, Oregon, built in 2000 for $130 per sq. ft., to a speculative small business park office building (about 65,000 sq. ft.) in the Kansas City, Kansas, area, built in 2002 for under $90. In each of these cases, building developers were convinced that they would be better off long-term with a fully documented and certified project. In the Oregon project, the owner had a personal commitment to the environment and wanted to demonstrate it with this project. In the Kansas project, the owner anticipated that the LEED certification publicity would help him find tenants who had similar environmental concerns, and he was right. *In a highly competitive market for office space, particularly in suburban areas, often a slight edge will translate into a market decision for your offering and against that of a competitor.*

In the case of government buildings, there has been substantial acceptance of LEED as a standard for both developing better buildings as well as demonstrating public commitment to higher levels of environmental responsibility. For example, the city of Seattle, Washington, adopted a policy in 2001 that all new public buildings over 5,000 sq. ft. would have to be LEED Silver-certified. States are trying "performance-based" LEED contracting, as they strive to meet their real estate needs without putting out the upfront capital. In these situations, the agencies are asking for guarantees of specific LEED achievement levels from private developers, typically LEED Silver, who often employ a design/build project delivery method. Such projects offer significant marketing opportunities for design/build teams which really understand the LEED system and the costs of attaining various levels of certification.

Given the resistance of many owners and developers to undertaking the costs and uncertainties of LEED certification for commercial and institutional buildings, it is important for marketers to have another design approach that can be put into place immediately, either in conjunction with LEED or as a "stand alone" integrated design tool, to deliver "best in class" high-performance buildings with the design professionals you are comfortable in using for your projects. As one such tool, the E-Benchmark guidelines for new construction (www.poweryourdesign.com) provide detailed design guidance for the 15 major climatic regions of the U.S., from dry to humid and hot to cold; in this sense it is more detailed and "prescriptive" than the LEED performance-based standard, but probably easier to approach for most mechanical engineers and architects, in that it "tells you what to do" in most cases to achieve a given result.

What are high-performance buildings? What are their design and operating characteristics? How can marketers translate these characteristics into benefits that building owners and developers will value?

- **They save 25% to 40% or more of a typical building's energy use** by incorporating high-efficiency systems and conservation measures in the basic building envelope, HVAC plant and lighting systems.

- **They are "commissioned," through the use of performance testing and verification** before the end of construction, for all key energy-using and water-using systems. *Often, the documentation provided by the commissioning process can be helpful later on in troubleshooting problems with building operations.*

- Green building commissioning involves **third-party peer reviews during design**, to confirm that design intent has actually been realized in the systems chosen. Systems such as the Advanced

Building guidelines provide a clear definition of various levels of building commissioning, so that a building owner or developer can choose the approach that best suits their situation and get it priced for the actual services required. Some programs also include an "end of warranty period" or "11-month" spot check on all key systems and preparation of a manual for "re-commissioning" the systems at some set interval, typically five years.

- They achieve **higher levels of indoor air quality** through a careful choice of paints, sealants, adhesives, carpets, and coatings for the base building and tenant improvements, often in conjunction with building systems that provide higher levels of air filtration and carbon dioxide monitors to regulate ventilation according to occupancy. With so many building occupants these days having breathing problems and chemical sensitivity, it just makes good business sense to provide a healthy building. Documentation of these measures can often help provide extra backup when fighting claims of "sick building syndrome."

- They often **incorporate daylighting and views to the outdoors**, not only for occupant comfort, health and productivity gains, but also to reduce energy costs. There is a growing body of evidence that daylighting, operable windows and views to the outdoors can increase productivity from 5% to 15% and reduce illness, absenteeism and employee turnover for many companies. Throw in higher levels of building controls that allow for such things as carbon dioxide monitoring and ventilation adjustments, for example, and one has an effective program for the "people problem" that can be sold to prospective tenants and other stakeholders. *For owner-occupied buildings, these savings alone are often enough to justify the extra costs of such projects*. Considering that 60% or more of the operating costs of service companies (which most businesses are) relate to employee salaries and benefits, it just makes good business sense to pay attention to productivity, comfort and health in building design and operations.

From a marketer's standpoint, what you want to sell is "the sizzle, not the steak." So you really need to understand how owners and developers see the benefits of these buildings.

THE PROCESS FOR CREATING A GREEN BUILDING

Conventional Design Process

Often, the conventional "design-bid-build" process of project delivery works against the development of energy-efficient and green buildings. In this process, there is often a sequential "handoff" between the architect and the building engineers, then to the contractors, so that there is no "feedback loop" arising from the engineering design, to building operating costs and comfort considerations, then back to basic building design features such as glazing, envelope, orientation, structural materials and building mass.

In a more conventional design process, for example, the mechanical engineer is often insulated from the architect's building envelope design considerations, yet that set of decisions is often critical in determining the size (and cost) of the HVAC plant, which can often consume 15% to 20% of more of a building's cost. As a result, key design decisions are often made without considering long-term operations. These decisions often result in higher costs and lower operating efficiencies for building owners and developers. Often adding some extra cost to the building envelope, through improved glazing and other solar control measures, can reduce the HVAC system costs by far more, thus freeing up funds for further improvements, in what can be described as a "virtuous cycle."

Integrated Design Process

As we said earlier, most developers and designers find that the process for creating green buildings requires an "integrated design" effort in which all key players work together from the beginning. (See Figure 8-1 below for an approach to diagramming the integrated design process). Developers and owners have discovered cost savings of 1% to 3% (of initial budgeted capital costs) in building design and construction through the use of integrated design approaches as well as other "non-traditional" measures, which might include bringing in the general contractor and key subcontractors to help with pricing alternative approaches to comfort in a building. There can also be time savings as well: considering all design elements upfront often prevents costly and time-consuming "re-design" after "value engineering" has attacked the first design in an effort to meet changes in budgets or project requirements later in the design process.

Marketing green buildings to owners and developers involves having a good grasp of the costs and benefits associated with the integrated design approach, since very few clients have had the experience of a completely finished green building project at this time, using this new design approach. Such an approach often requires greater design fees for which the owner hopes to make up in lower construction costs. From a marketing standpoint, firms might want to take a "risk sharing" approach, in which a portion of the design fees are "performance-based," and are paid upon achieving specified modeled levels of energy efficiency. This puts the onus particularly on the architect and mechanical engineer to work closely together to integrate decisions involving the building envelope with those involving lighting, daylighting, comfort and the HVAC system.

Marketing integrated design needs to begin when the team is being interviewed for the project assignment. During the interview process, each team is often asked to be specific about their approach to the forthcoming project. For green buildings, the integrated design approach often includes holding "eco-charrettes" or "sustainability forums" with key non-technical stakeholders during programming or conceptual design, as well as an "eco-charrette" with key design team members at the outset of schematic design. These charrettes are often an economical and fast way to explore design options as a group and all at once, before settling on a preferred direction. In the charrettes, everyone gets to provide input on building design before design direction is "set in stone." Often these charrettes are facilitated by an outside, well credentialed third party, providing a marketing opportunity for additional professional services[51].

However, integrated design approaches often involve greater upfront costs and time allocations than conventional building programs. In negotiating fees for its work, the architect often needs to re-educate the client about the value of this approach, and to get money (and schedule time) to carry it out. This negotiation is critical to the final outcome of a green building project and needs to be thought about as a continuation of the marketing effort.

Considering the effect of design process, one report states,

"We found successful projects--ones that achieve a high LEED score and stay within their original budgets--are ones where the design team sits down with the owner right at the beginning to talk about sustainable design and clarify goals. [That way] everyone on the team has some input. The most successful projects have a very integrated process. The projects where it's not working as well is where some member of the design team takes on [the LEED elements], but is doing it separately from the rest of the team."[52]

While the integrated design process sounds simple, it goes against much of current design practice and requires architects, engineers and designers to develop a different process and set of communications skills for handling the initial stages of a project, as shown below in Figure 8-1.

PART THREE

CASE STUDY—MARKETING SOLAR ENERGY SYSTEMS

Market transformation for solar energy systems is gaining increasing importance as we move through the second decade of green building practice (using the formation of the USGBC in 1993 as a starting point). A recent survey illustrates the opportunities and challenges facing marketers for solar energy products and systems in commercial and institutional projects. The new federal tax credits of 30% for commercial building solar systems deployed in 2006 and 2007 will give new impetus to the marketing of solar power.

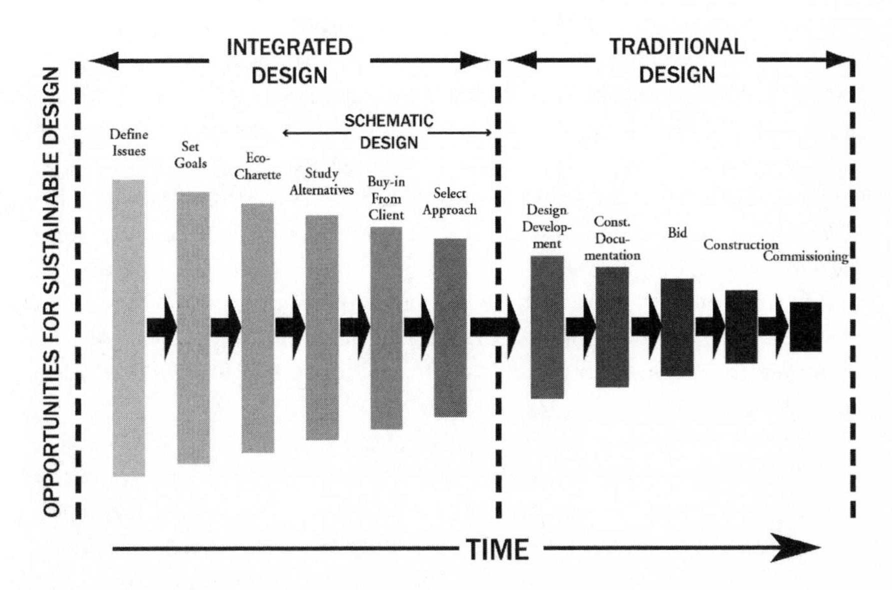

Figure 8-1. Integrated Design Process, Showing "Front Loading" of Design Effort vs. more traditional "back loading"

Survey Design

In May of 2004, the author conducted a survey of nearly 1,000 building industry professionals in his professional database, using a web-based survey tool and a 20-question survey instrument. We eventually received 223 responses or about 22% of the total surveyed population.[53]

Survey Participants

Survey participants came from a range of disciplines and occupations, including 47% architects, and 22% other design team members (typically engineers) and contractors. So, about two-thirds of survey respondents were directly involved in building design and construction. In terms of interest and experience, 18% had already done a LEED-certified project, and 25% were doing a LEED-registered project at the time of the survey. Another 31% were doing projects with sustainability goals (but not LEED-registered), so that nearly 75% of survey respondents were clearly interested in LEED or sustainable design.

Survey Results

Of the survey participants, 56% were working on green office buildings, 35% on higher education facilities, 27% on cultural and recreational projects, and 25% on K-12 school projects. When asked if they had considered using solar energy in any of their projects, 84% said "yes," with 73% considering photovoltaics (PV) including 51% with building-integrated PV, 57% solar water heating and 19% solar pool heating. Of these respondents, 59% currently have a project in design, 28% have at least one project in construction and 26% have an operational project. This indicates that firms designing for PV or solar thermal applications tend to do more than one project, as their design, construction and operational experience grows. We should note that only 16% of those who considered a PV or solar thermal project ultimately decided not to proceed with it. (This means that once solar gets "on the table," a project is likely to go ahead with it.)

Survey participants who decided not to proceed with a solar project overwhelmingly (55%) said cost was too high and (52%) the payback period was too long. The plain fact is that most solar applications (even in the Southwest) cannot compete with other building energy-efficiency measures that have a much higher economic return. Less than 10% said that they didn't have proper solar exposure or that there were design considerations that prevented the use of solar. Since a 100-kW solar system costing $700,000 or so to install (without considering tax incentives or rebates) will produce less than 200,000 kWh of electricity per year in most U.S. locations (valued at $12,000 to $30,000 in most utility service areas), it is not surprising that initial cost is the major barrier to more widespread solar adoption.

To facilitate marketing as the solar power industry grows, firms should focus on developing expertise, project experience and a recognizable "name" in the early stages of this new market. There is an important role that developing internal firm expertise plays in convincing building professionals to "stick their necks out" and become advocates. Many professionals still give more weight to first cost increases, find LEED projects harder to justify, and say that the market was unwilling to pay a premium for sustainable design. On this basis, it is hard to justify LEED certification to one's clients, and one finds that the market is uncomfortable with innovative sustainable design measures such as solar thermal and PV systems. This suggests in some ways that incorporating solar design measures and integrated design approaches into the basic practice of a firm ("if you hire us, you get the following green measures, no discussion, no argument" approach) might be more effective to help firms differentiate themselves in the marketplace.

Many engineers are more comfortable with designing solar systems when there are financial incentives (to cut initial cost), a client insistent on using PV or solar thermal (perhaps because it is so visible and because most people would recognize a solar power system without being told) and perhaps dramatic increases in local electric utility rates (a repeat of the artificial electric power "crisis" in the summer of 2000 and 2001). While the return on investment for solar projects may be fairly good for private owners in some states, the intangible "PR" benefits of a visible green measure are also significant. The new Energy Policy Act of 2005 (EPACT), with its 30% federal tax credit for commercial solar thermal and PV systems, may cause more private projects to take this approach, owing to the implicit support of the federal government.

At this time, LEED has gained perhaps 10% or more of the institutional market for new buildings but scarcely 3% of the corporate market[54]. So, for the private sector market, the clients can be described as "innovators" and for the public buildings market, the client base is more likely of the "early adopter" category. Even in the public buildings client base, many project managers who supervise large projects could properly be characterized as late adopters, and will need strong mandates from upper management to accept sustainable design projects or expensive solar systems, especially since most building projects have constrained budgets, as prices for concrete and structural steel, the two main commercial building materials, have skyrocketed in the past few years.

Anecdotal evidence of overall benefits favors solar power, but it has not filtered yet into the general marketplace enough to overcome perceived cost hurdles. Since most green building markets are "project based," it may take some time for perceived benefits to find appropriate projects, for a fuller implementation. Oftentimes, adoption of innovation is incomplete, for example, when a technology is desired (in the way of desired outcomes such as LEED certification or PV use) but not deployed into general use; this phenomenon has been called the "acquisition gap" and has been found in a number of technology diffusion studies, wherein the authors observe that "knowledge barriers impede deployment.[55]" Therefore, green building marketers interested in promoting solar in their projects would do well to spend time educating the client on the multiple benefits of such systems, preferably early in the design process.

In the light of the current state of the market, our survey respondents' desires for more independent cost and performance evaluations of solar power systems are critical for building credibility and overcoming perceived barriers. In my own professional experience, the expectation of real benefits has to exceed the likelihood of increased costs by 25% or more ("Yudelson's Law") to change most decisions in favor of new technologies or methods. As a technology or approach such as LEED moves more into the mainstream, it is more likely to meet with this type of resistance. Many studies of the psychology of decision-making have shown that consumers and clients are likely to resist change unless they perceive the "downside" risk to be heavily outweighed by a well documented "upside" benefit.

The Current Market for Solar PV in Buildings

The installation of solar PV in grid-connected commercial and industrial applications in 2004 was about 61,000 kW, valued at about $400 million and representing nearly 50% of the total U.S. installed solar power that year. Installed solar PV applications in distributed grid-connected applications have grown nearly 1000% since 2000 and are expected to exceed 100,000 kW in 2005.[56] This rapid growth augurs well for PV applications in commercial, institutional and industrial buildings, as costs are coming down and experience with designing and specifying the technology is growing. (The number of total installations in the U.S.

represents only about 10% of the world's use of PV technology in 2003). Residential grid-connected uses accounted for only about 4,000 kW in 2003, about 6% of the total U.S. usage. (If a typical consumer application is 2 kW, then there were about 2,000 grid-connected systems installed in 2003, still a very small percentage of the annual home-building activity of nearly 1.8 million residential units).

The Economics of Solar Power in Buildings

To be honest, there is no compelling economic case for including solar energy systems in commercial and institutional projects, such as there is for energy efficiency measures, daylighting, passive solar design, and similar measures. Let's take a look at the economics of solar power, for projects built in the U.S. in 2006, using the author's analysis in Table 8-3.

One could conclude that right now, in Oregon and other states with generous PV incentives, *there is a reasonable economic case for private-sector projects* to consider using solar electric technology. Note that the return on investment is based on current power prices; the actual economic benefits might be greater if peak power prices are much higher, if base power prices are higher than $0.10 per kWh, and might be much less if annual maintenance costs are more than zero!

Mainstreaming Solar Technology

If solar building technology is to enter the mainstream represented by the "early majority," it must begin to take note of the problems of marketing new technology well illustrated in the classic "Crossing the Chasm," in which Geoffrey Moore demonstrates how difficult it is to go beyond the early adopters to the more general marketplace, using the same marketing mechanisms and communications tools as for the smaller, more specialized and less-risk-averse group of innovators[57]. In other words, the mainstream market and the "gearhead" market require totally different marketing approaches and communications channels. For solar power systems, we would argue that they need to be packaged in standard modules, not requiring any advanced design engineering, employing more of a "plug and play" solution. For commercial systems, this might include putting inverters and all other electronics on the roof of a building, with simple connections to the building's electric power system or even directly running DC lamps and other related equipment from the PV system (to reduce efficiency losses through inverters converting DC power output to AC power.)

Table 8-3. Economics of Solar Electric Power Use in Buildings

Basic cost of the system:	$8,000 per peak KW (Kilowatt); can be less for larger systems.
Power output:	1,500 kilowatt-hours (kwh) per year per kW of rated power (average for a good part of the United States, lower in the Pacific Northwest – 1100 kwh net - and maritime Northeast.)
Value of energy produced:	@ $0.10 per kwh, $150 per KW (peak); at $0.20 per kwh, $300/KW (peak) (assuming all power produced can be sold back to the local utility at full retail rates)
Return on investment (ROI):	$150/$8000 = 1.875% (excluding annual maintenance costs), or 3.75% at the higher rate of $0.20/kwh.
Potential subsidies:	30% federal tax credit (in 2006 and 2007), state support in California at up to $2.80 per watt; other incentives may vary. Federal depreciation credits may also apply for profit-making entities. Oregon has Business Energy Tax Credit valued at about 25% of initial cost and utility payment of $0.15 per kwh produced, from the Energy Trust of Oregon.
Tax-paying (Oregon) ROI:	6%, assuming $0.15 per kwh value of power, net federal depreciation present value of 25% of cost; Business Energy Tax Credit, net present value @ 25%. (note state tax credit reduces federal tax deductions for state taxes paid, by the amount of the effective tax rate); 30% federal tax credit. Assumes lower PV output at Portland.
Tax-paying (California) ROI:	6%, assuming $2.80 per watt state grant (35% of cost), net federal depreciation present value of 25%, 30% federal tax credit, no state tax credits, and power valued at $0.10 per kwh. ($150 value for power, $2,560 net system cost).

Malcolm Gladwell's work discusses how epidemics and fads spread, a topic of great relevance to the diffusion of innovation, especially in the areas of understanding the roles of communications channels and social networks[58]. Basically he posits that innovations spread fastest through the work of a relatively few people who have well-developed social networks; when they are "sticky" in terms of the emotional effect of memory, myth and metaphor; and when disseminated within a powerful context (almost a tribal setting) by people others know, trust and like.

In Gladwell's terminology, green building and solar energy adoption will spread most rapidly through the actions of well-connected individuals (word of mouth spreads most fads); through people who widely and openly share their knowledge with others (mavens or experts whose judgment is trusted); and through "persuaders" who have the ability to tell compelling stories to others. Imagine the American Revolution occurring without Paul Revere (the connector) and Tom Paine (the persuader). In other words, innovations finally spread when good salespeople get involved. Green building and solar power have the first two categories in abundance, but the third is scarce.

The emotional appeal of widespread solar energy adoption in American homes and businesses might be an unexpected consequence of the current war in Iraq and the resurgence of oil prices above $50 per barrel in 2005. If the American public will finally wakes up to the true costs of the current energy dependence on oil imports it may determine for the first time in 25 years to do something personal about it. (Witness the overwhelming demand for hybrid autos beginning in 2003.) Solar power solutions are well positioned to take advantage of these trends.

Past Experience with Marketing Solar Energy Systems

In the author's personal experience, as a state official, lobbyist and marketer watching and participating in the diffusion of residential solar water heating technology in California from the period of 1977 through 1985, in spite of awesome tax and energy saving advantages and a relatively simple technology, *it was not until major sales organizations became involved that technology adoption accelerated*. In other words, most people were not picking up the phone and trying to buy a solar water heater; they were waiting to be sold. Imagine the automobile industry succeeding without sales-oriented local dealerships. The difference is critical: in most surveys I've seen and conducted, most people are waiting for "someone else" to take the lead in green buildings.

For example, in 2003 did architects begin to recognize that building-integrated PV systems, for example, can be part of a significant architectural statement, with projects such as Colorado Court in the Los Angeles area, a $4.2 million, 30,000 sq. ft. apartment project, which won a national AIA award for a five-story high wall of 200 PV panels[59].

Recommendations for Solar Power Marketers

In the marketplace for solar PV systems, marketers need to push their companies for the following type of information:

- Case study data, with solid cost information, including initial cost increases. This means widely publicized data, by region, based on actual project costs.

Colorado Court, Santa Monica, CA, ***Image Courtesy of Pugh & Scarpa Architects***

- Comparative cost information within and across building types, as to the full costs of solar power applications, the resulting benefits, and ancillary features such as public education. Solid, measured performance data, in the field, will also be as necessary as cost data, to encourage trial by "early adopters."

- Demonstrable information on the benefits of solar power systems beyond well-documented operating cost savings from energy conservation. Such benefits might include better public relations, more newspaper and media articles (yes, large PV systems are still novel in most areas), more responsiveness to stakeholders (such as "walking the talk" for a firm committed to sustainable practices), and so on.

- Personal stories, by both practitioners and building owners, about the costs and barriers to completing projects with solar energy systems/applications.

- Stronger use of multi-media approaches and other modern sales tools, to increase the emotional "bonding" with solar goals and methods on the part of stakeholders and final decision-makers. One of the tactics I have explored with several clients is to "sell" the PV panels, one by one, in the manner of theater seats, to local stakeholders. This might work especially well for schools and nonprofits, which often seek ways to bond the community to their projects. For example, a local utility (electric or water) could charge $5/month for five years ($300 total), enough for a family to buy a PV panel for a school or public project.

Strategic Considerations

To summarize, solar power marketers need to understand how their marketing approaches must evolve in order to compete effectively:

- They must pick a strategy that incorporates high levels of differentiation or low cost, with explicit focus on particular market segments, that might include geographic, project type, owner type, psychographic profile (e.g., early adopter, early majority), project size or even technological approach.

- This strategy must be reinforced to become recognizable as a "brand identity" of the firm and its specific products or services. Internal reinforcement includes training, certification and notoriety as LEED practitioners and solar experts; external reinforcement includes activities to increase the visibility of the firm and its key professionals, including speaking, lecturing, networking and publicity for successful projects.

- Companies should consider developing their own proprietary tools, as part of a branding approach. Firms should also develop methods to execute solar projects with modest additional design fees and to utilize all available state, federal and utility incentives for solar power applications.

- Architects and engineers must form closer working alliances with contractors and other project professionals to ensure that their solar power designs can actually get built within prevailing project budget, time, technology, expertise and resource constraints.

- Designers should look for opportunities to level the playing field for solar power by incorporating building-integrated PV into the next project they design; BIPV systems substantially change the economics of solar power by offsetting some of the building's expensive "skin" costs ($60 to $100 per sq. ft.) with solar panels costing $50 to $100 per sq. ft. They also offer a wide range of colors and aesthetic possibilities in building design that would make the energy production a "bonus" feature.

The author's own professional experience suggest that solar power advocates and building design professionals need to become aware of the theory of innovation diffusion and strategies for creating competitive advantage, if they are to successfully spread the message of solar power beyond the current group of "early adopters," who are the primary market at this time in the institutional and governmental sector and the "innovators" who still dominate the green building market in the for-profit sector. (See Chapter 10.)

Chapter 9

Understanding Marketing Strategies

WHO IS USING LEED?

The USGBC has documented the uses of LEED by public, private and non-profit organizations. As of the end of September, 2005, the number of LEED registered projects could be categorized by end-user as follows:

- Corporate 26% (the "for profit" market)
- Local Government 23%
- Non-profit 20%
- State Government 12%
- Federal Government 9%
- Other/Individual 10%

It turns out that the *total building area* of LEED projects represented by the corporate sector is about 35%, owing to the larger average size of those projects. For the same reason, the federal share of building area (and typically overall market size) is also a few percentage points greater as a percentage of the total.

These data show that government agency buildings represent nearly half the total projects, with government and non-profit corporations together comprising about two-thirds of all projects. Corporate projects tend to be larger projects (typically for major corporations), with a smattering of local small companies with significant environmental goals or missions. *For marketers, the clear focus at this time has to be governmental and institutional projects, if their firms have experience in these sectors.* This conclusion is reinforced by economic difficulties and oversupply of office space in a number of major markets that have slowed commercial building activity during 2004 and 2005 and made it even more sensitive to initial cost.

Another way to look at the LEED registered projects is by end-use. Just about every conceivable project type has been LEED registered, including an Oregon wine-making facility, mostly underground! What marketers should understand is that many public projects are likely to carry requirements either for a firm's having either LEED project experience or LEED APs on staff. Large adopters of LEED such as the federal government are beginning to consider having LEED-registered projects as 10% or more of the evaluation of a prime contractor's qualifications.

Given that it often takes two years or more for projects to move from design to completion (and certification can only take place after substantial completion of a project), marketers should be pressuring their firms and their clients now to step up and participate in the certification of existing or upcoming projects in 2006. Some firms are even taking the extraordinary step of providing the LEED project certification documentation (which can take from 100 to 200 hours of professional time) "pro bono" to valued clients, just to make sure that they can certify the project and have at least one on their resume. Considered as a marketing expense, such pro bono time is not large in the overall marketing budget of mid-size (30 to 50 people) or larger firms.

The first 195 LEED version 2.0/2.1 certifications had achieved the following levels. There were also 19 LEED v. 1.0 certified pilot projects, but since these are "ancient history," we do not deal with them in our analysis, although firms certainly do include them on their resumes.

- Certified: 88 projects (45%)
- Silver: 60 projects (31%)
- Gold: 42 projects (22%)
- Platinum: 5 projects (3%)

The 42 LEED Gold project certifications have included such varied building types as:

- Renovation of a 100-year old warehouse in Portland, Oregon
- A developer-driven technology park in Victoria, British Columbia
- An elementary school in North Carolina
- An office/warehouse building in Gresham, Oregon
- A non-profit office building in Menlo Park, California

- Two projects for Herman Miller Company in Zeeland, Michigan
- A public office building leased to the Commonwealth of Pennsylvania
- A very large state office building in Sacramento, California
- An environmental learning center in the Seattle, Washington area

Of the 42 LEED-Gold projects, 13 (31%) are corporate projects, while the balance are public agency, educational and nonprofit in nature. This reflects the split of such projects mentioned above and suggests that the most immediate impact of LEED will be on those firms that market to the public sector.

ROLE OF BUILDING PROFESSIONALS

Each professional discipline has a role to play in a typical building project.

- **Architects** naturally have the task of coordinating overall building design and of dealing directly with the building envelope, daylighting, materials selection, window and roof specification, etc.

- **Interior designers** have to deal with materials selection for furniture and furnishings, and to help specify low-VOC paints, carpets and similar low-toxicity items. They may also be asked to assist with specifying elements of underfloor air distribution systems, such as carpet tile.

- **Mechanical and electrical engineers** can contribute between 25% and 50% of the total points required for LEED certification, focusing on water use, rainwater reclamation and gray water reuse systems, energy efficiency, lighting design, commissioning, indoor air quality, carbon dioxide monitoring and thermal comfort.

- **Energy engineers** are called upon to prepare energy models for buildings including Computational Fluid Dynamics (CFD), and to design on-site power systems such as microturbines and combined heat and power (CHP) plants.

- **Civil engineers** have to deal with stormwater management, provide input on rainwater reclamation systems, prepare erosion and sedimentation control plans, and sometimes advise on constructed wetlands, bioswales and on-site waste treatment systems.

- **Landscape architects** need to consider water efficiency of landscaping design, input to design of detention ponds, bioswales and constructed wetlands, and also oversee site restoration programs.

- **Structural engineers** are asked to consider the relative benefits of wood, steel and concrete in structural systems, given their different effects on sustainable design. Often projects that use passive thermal conditioning require heavy mass structural components such as concrete. Structural engineers also have a role to play in green roof technology, since weight is added to the structure.

- **Cost consultants** have a significant role to play in assessing the costs of innovative green building systems, such as "eco-roofs," solar power, and stormwater retention systems, as well as advising clients on the overall costs of green buildings.

- **General contractors** have to provide for recycling of construction debris (often at a 90% or better level) and of documenting the costs of all of the materials that go into a building. They oversee the construction indoor air quality management plans and activities, and they play a vital role in documenting the costs of the project. Contractors are also responsible for construction staging (LEED Sustainable Sites credit 5.1) and erosion control plans.

- **Subcontractors** are often asked to work with unfamiliar or hard-to-obtain recycled content materials and to document the costs they incur. Mechanical and electrical subcontractors often have to interact with the building commissioning process as well.

- **Environmental consulting** firms also have a role to play in sustainable site selection practices and assessment of the potential for on-site storm water management, brownfield redevelopment and site restoration, for example.

GREENING A DESIGN FIRM

Building sustainable design capabilities at architectural and engineering firms engaged in green buildings can take many forms. An in-depth survey of 20 companies by *Environmental Building News* was reported in May of 2004.[60] Firms reported six major areas of activity, similar to our own survey data from hundreds of firms reported elsewhere in this book.

- An in-house Green Team can offer internal consulting to projects.
- Internal training and education, including staff-led and vendor-led in-house sessions and support for attending conferences and outside trainings.
- Management of green building information, including a library and development of in-house specifications for green projects.
- Tools for designers to use, including energy modeling tools and metrics for determining "shades of green," such as LEED.
- Include expertise from outside (this is one of the most-effective, but least favorite measures, in my experience, owing to cost and the perception that "we can do it ourselves"), or use capable sub-consultants for projects (in the case of architects, this would include mechanical, electrical and civil engineers).
- Set goals for green projects, including LEED for client projects and internal assessments using LEED for all projects. Some firms start every project with an intent to "green" it as much as possible, regardless of budget or expressed client interest.

ASSESSING SUSTAINABILITY MARKETING STRATEGY

Certainly for most firms, the key marketing strategy of our time is "focus and differentiate." Most firms know their areas of focus fairly well, so the issue becomes how to differentiate a firm's capabilities in sustainable design from other firms'. Here are some suggestions for marketers of design and construction services.

Strategic Assessment

Often a firm needs first to conduct a strategic review of its capabilities and opportunities using a "SWOT" (Strengths, Weaknesses, Opportunities, Threats) analysis, a well-known tool for assessing the following areas of concern. See Figure 9-1.

- **Strengths** (internal, including staff skills, project history, client relations, cost structure, competitive position within its market sectors, knowledge of green design, interest in green design, knowledge of specific building types, financial strength, etc.)

- **Weaknesses** (internal, typically include lack of experience with green design projects, strong local and regional competitors who are advanced in such experience, lack of resources to hire the people the firm needs to buttress its expertise, etc.)

- **Opportunities** (external, including market trends, growth in various market sectors, new laws and regulations favoring green buildings, new financial incentives for green buildings, actions of competitors, industry dynamics, profitability of various market segments, new developments in green technology, etc.)

- **Threats** (external, changes in client policies to favor firms with green design expertise and completed projects, stronger competitors opening offices in a firm's home markets, etc.)

Based on this review, a firm can much better assess its areas of maximum opportunity and direct its marketing efforts in the most cost-effective manner.

INTEGRATING SUSTAINABLE DESIGN INTO MARKETING FOR A DESIGN FIRM

How should marketers advise their firms to take advantage of market opportunities in sustainable design? Here are some of the methods various firms have found successful.

Make a Major Firm Commitment to Sustainability

Through its membership in the Oregon Natural Step Network and

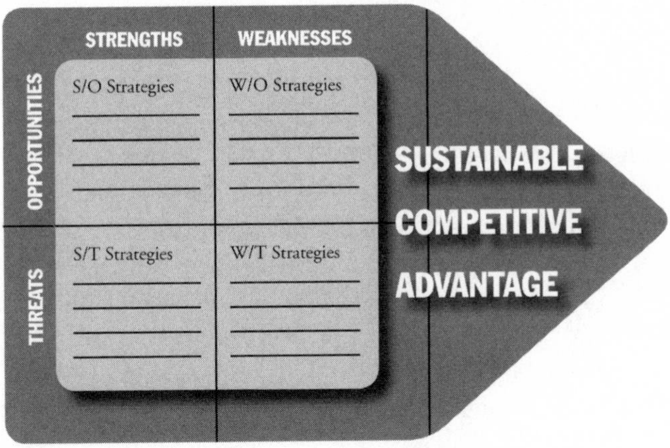

Figure 9-1. Components of a SWOT Analysis

its active participation in the U.S. Green Building Council (the author is a former national Board member and a LEED Workshop national trainer while serving as Sustainability Director for the firm), Interface Engineering started on the pathway to corporate sustainability in 2000. Other firms have gone even farther: one Portland architectural firm has three internal committees that address first, sustainability at home (for all firm members); second, building up the firm's internal sustainability activities; and third, examining each project for its success in incorporating sustainable design elements (case study of BOORA Architects, found at www.ortns.org). Still other firms have hired sustainability coordinators to set up and manage internal information and to provide expertise and resources to each project. Other firms have set up separate internal profit centers to offer their sustainability expertise as consultants to both their own projects and to external clients. Finally, some architectural and engineering firms have taken advantage of planned moves of their own offices to experiment with green design, participate in LEED-CI pilot projects and to show everyone that they can "walk the talk."

Interface Engineering is a four-office firm based in Portland, Oregon (www.ieice.com), serving projects primarily in Washington, Oregon and California, but with experience in more than 40 states and several foreign countries. The firm employs about 110 people and ranked 56th nationally among similar mechanical and electrical engineering firms, in terms of 2004 revenues.

At Interface Engineering, some activities to promote sustainability include:

- **Internal education**: the first step is to put well-trained and knowledgeable people onto project teams. The firm is complementing its normal continuing education in energy engineering, lighting design, plumbing engineering and related topics, with a strong in-house training program in the LEED green building rating system. The firm has begun training its corporate staff in the Natural Step system and has taken steps to use Honda Civic hybrid cars for project travel. At September 2005, the firm had 25 LEED Accredited Professionals among its technical staff of over 90 people.

- **External education**: To build a strong sustainability presence in the marketplace, Interface Engineering offers its expertise in public and private seminars to clients such as architects and owners. For several years, the firm has been holding breakfast and lunch seminars in sustainable design for clients and fellow consultants, and offering public seminars as part of a group of AIA-approved continuing education courses for architects.

Publishing and Speaking

Over the past five years, Interface's internal experts have published articles on water conservation, energy engineering and sustainable design in several major national trade publications for mechanical and electrical engineers (*Consulting-Specifying Engineer* and *HPAC Engineering*), in magazines for the design profession (*Environmental Design & Construction*), as well as articles in local design and construction industry media. Sustainability experts from the firm have spoken to college and university architects, planners and facility managers, as well as at national and international green building conferences. This intensive commitment to external communications lends credibility to a firm's claims of expertise and also provides reprints and other opportunities for marketing this expertise.

Sustainable Improvements in Operations

Interface Engineering completed a major headquarters move from the suburbs to downtown Portland at the end of 2002. As part of this move, the company was able to increase its commitment to sustainable building operations, including extensive daylighting and healthier indoor environmental quality for its own people. In addition, there has been a documented increase in the use of public transportation, with the total number of participants taking bus or light rail, bicycling and walking to work, rising from less than 10% to more than 60% of staff. The firm has also recently moved to increase its paper recycling by taking away "trash" cans from each employee's workstation and substituting paper recycling boxes instead. Finally, the firm purchased Honda Civic hybrid vehicles for its small fleet to reduce consumption of gasoline, and it is passing the savings along to clients with lower charges for traveling to meetings.

Sustainability is not a destination, but a journey. By making a strong company commitment to sustainable design and operations, many firms are beginning to "walk the talk," in an open way. Clients appreciate working with firms that share their values and that are willing to experiment with new technologies and processes. This is true contemporary marketing: *building relationships based on shared values.*

Capabilities

Know what your firm principals and senior level personnel are doing in the area of sustainable design and learn what they are hearing about the need for these services among your client base. Incorporate all sustainable design projects into the firm's standard capabilities statements (SOQs) and proposals. (Many projects have sustainable design elements that can be used without necessarily being LEED-registered; the author's estimate is that perhaps only 30% of the projects with sustainability goals ever register with LEED, owing to cost considerations.) Make sure you're familiar with the language of sustainable design for your professional area and, if you're the firm's chief marketer, push the technical types to "clue you in" where your own knowledge base might be a little weak.

Competitors

Know the strengths and weaknesses of the competition in this area of design and construction, so that you'll be prepared to match their strengths and exploit their weaknesses in the proposal and interview stage. You may even decide not to respond to a solicitation from a client asking for sustainable design, if you think your firm can't yet stand up to the competition for a certain project type or for a client that is already experienced in LEED projects.

Differentiate

Make sure clients know how your firm will approach the project differently from major competitors by showing your team's design tools and understanding of sustainable design. One North American mechanical engineering firm has shown its commitment to the LEED system, for example, by certifying more than 60% of its staff as LEED Accredited Professionals, including some not directly involved in design, and by eagerly embracing and introducing new technologies in its area of expertise. As a result, this firm has established firm roots in new geographic territories with innovative green architects and is well positioned to make further inroads in a wider geographic area. (How will YOU compete against such a firm, if they decide to expand into your neighborhood?) In a 2002 article, "Post-Modern Engineering," I outlined steps that engineering firms should take to position themselves as innovators attuned to the new sustainable design paradigm. (See Chapter 13 for an updated version of this article).

People

Make sure that a large number of your firm's key people become LEED Accredited Professionals (LEED APs). With more than 21,000 accredited professionals nationwide, there is no longer any excuse for a firm not to have several on its staff. Public LEED-NC workshops are offered nearly 50 times a year by the USGBC, so there's bound to be an intensive training somewhere close-by to which you can send key people. LEED-EB and LEED-CI workshops are also offered nearly 25 times a year somewhere in the U.S. The largest U.S. architectural firm, Gensler, has more than 400 LEED APs, as of 2005, and many other major firms are adding to their LEED AP totals (see p. 11).

Case Studies

Build a portfolio of LEED registered and LEED certified projects as quickly as possible. Look for other projects that have sustainability elements and try to incorporate them into your case studies as quickly as possible. Interface Engineering built a library of case studies on successful project experience using sustainable design that it uses to market these services and provide to the media to help in profiling the firm's expertise.

Press

Tell your story aggressively to as many media outlets as you can.

Successful sustainable design projects are still rare enough in many areas of the country and in specialized market niches (even large market segments, such as K12 schools, only have eight of the first 106 LEED version 2.0 certified projects) even rarer. Publications in all vertical markets are publishing articles on sustainable projects on a regular basis. These are one of the main vehicles for new clients to become aware of your firm.

USGBC Membership

Membership "has its privileges," to borrow a phrase. Make sure your company joins the U.S. Green Building Council and can use its logo on proposals, stationery and brochures. Joining the USGBC will signal to clients that you have the interest and knowledge they are seeking. The cost will range from a few hundred dollars up to $2,500 a year for fairly large professional service firms. This is probably the best investment a firm can make to establish credibility with clients. Join the local USGBC chapter and become active in it. This is ideal networking territory.

External Marketing

It's essential for your principals and key staff to share their knowledge and enthusiasm for sustainable design with potential clients on a regular basis. You will find out what your clients know and want, and what your people don't know and should learn. Prepare to offer sustainability services as an "extra service" on all major proposals to your clients (but be prepared over time to have to include most of these design services in the base fee, as clients learn what is and isn't required for LEED projects). Be prepared to explain to them why this approach will not only benefit the project directly, but could also result in major marketing benefits for their project, company or organization. The author has always advocated sharing knowledge in the form of talks, articles, classes, seminars and one-on-one discussions; leading professional firms can successfully differentiate themselves by sharing knowledge with clients and the larger green building community in an appropriate way. This often leads to "casual marketing" through word-of-mouth referrals, improved relationships and team-building.

Focus

A final cautionary word: not every client is a candidate for green marketing. Not every client wants to be the "first kid on the block" to have a "new toy" or to be a technology leader. While many building owners and institutional facilities managers trust their architects and are willing to follow the architect's lead in pursuing a green building agenda (especially true in my experience for higher education projects), most corporate and building owners are more cautious, and speculative developers, for the most part, are still in the "wait and see" stage. So, focus most of your marketing efforts on the more adventurous owners.

Integrate Green Design and Marketing Activities

Once a firm secures a sustainable design project, the marketing work has just begun, for a successful effort is always the best marketing tool, and one cannot wait for a project to be finished (which might take two to three years) to start generating enthusiastic client support for referrals and testimonials. Early design activity, such as "eco-charrettes and "green forums," should also have as components a clear presentation of the areas of risk and ambiguity in the project and should develop explicit strategies for dealing with them.

These strategies might include:

- literature research and site visits to similar green projects
- early design modeling of daylighting, energy efficiency and natural ventilation opportunities
- early design interaction with materials and equipment vendors
- development of a design program that will not preclude effective green building measures.

(One project the author knows about had a client demand from the inception for air conditioning. This was in a very mild western coastal climate, and natural ventilation strategies were quite appropriate for the intended use. By giving in to this demand early, the designers have added cost to the project and precluded some more elegant design approaches to thermal comfort).

The theory of "diffusion of innovations" gives powerful insight into this behavior. (See Chapter 10 for a more detailed presentation). Only about 3% of clients are likely to be "innovators" and willing to pursue a new design trend or technology development before seeing how others have done with it. Another 13% or so are called "early adopters" who are likely following these trends and developments closely and are willing to try them once they see a few successful experiments or case studies. The remaining population of clients will not generally embrace change or take much risk, without clear evidence of benefit and a clear track record to examine. They are the "wait and see" crowd and, at this time, generally represent a waste of time for marketers.

This analysis suggests that architects and engineers need to be selective about which clients they pursue for green building projects and how they approach them. Your past successful (and documented) experience will be a powerful selling point in convincing clients to pursue LEED-registered projects with you. Additionally, designers should do research on other innovations the client has embraced in the past, what forces—internal and external—are driving the client to consider green design, and in which areas of technology and operations the client is likely to have greater tolerance for the risk and ambiguity inherent in taking new approaches.

CASE STUDY: MITHŪN ARCHITECTS, SEATTLE

Mithūn CEO Bert Gregory was interviewed in *The Marketer*, April 2004, the monthly magazine of the 5,000-member Society for Marketing Professional Services (www.smps.org), (for which the author served as guest editor[61]). As CEO and one associated with the firm for nearly 20 years, Gregory was instrumental in pushing it into a focus on sustainable design. The firm grew throughout the 2001-2003 national recession (gross service fees increased nearly 30% from 2002 to 2003) to about 145 staff members in 2004, including more than 40 LEED APs. It is widely seen as one of the leading green design firms in the Pacific Northwest.

When asked how the firm took a proactive approach to marketing sustainable design, Gregory stated:

> *One proactive element is our commitment to education. It is both internal to the team and external. Many of us speak locally or nationally about topics of sustainability…Those talks always have intangible but beneficial results in making people aware of our firm…Proactivity is making sure people are aware of us, making sure that we're establishing relationships and investing in our community.*

Sustainability has also changed Mithūn's practice of design by emphasizing collaborative strategies, using a broad consulting team at the start of a project. Gregory says that such strategies have:

Changed us to be in more of a leadership position on projects that are really complex and need lots of people to do them. More and more projects have an economist or real estate consultant as part of the team…These days we are spending more time sitting on the same side of the table as our clients, helping them understand the long-term economic impact, return on investment and choices they can make that will establish a higher value for their project or their portfolio. This is really different from how most architects would approach a project…The distinguishing feature of our practice has been our ability to incorporate design excellence with sustainable strategies.

Gregory believes the future of sustainable design is:

Really at the city level and at the broad-based master planning level…For individual buildings, the future is in clients and designers establishing goals that are extremely aggressive regarding environmental impact and understanding how we can truly create buildings with limited or no impact.

In terms of competitive posture, Gregory believes:

Ultimately, it [sustainability] is the cost of entry…For us, research and development is an important aspect of our practice. One way to continue to be a leader is to make sure you are doing R&D. Most strong businesses are including that in their practice.

In terms of actual practice, Mithūn has completed several LEED-certified projects[62] and many studies of urban sustainability, including two landmark studies of entire urban districts. The *Resource Guide for Sustainable Development in an Urban Environment*, focusing on the South Lake Union area in Seattle, is a landmark in green neighborhood design and can be downloaded from the Mithun web site.

Chapter 10

Understanding Segmentation, Targeting, Positioning and Differentiation

A marketer's job is always fraught with difficulty. How to make a "purple cow" (something remarkable) out of a "pink sow" (something ordinary) seems to be the perpetual task of the marketing arm of the firm. (In today's marketing environment, a firm must be "remarkable" just to get some attention, hence the "purple cow" analogy.)[63] In this chapter, we introduce some of the basic concepts of modern marketing and apply them to the issue of marketing green buildings, including design services, construction services, technologies and products. Segmentation, targeting and positioning are often referred to as the "STP" formula and form the essence of strategic marketing planning, as inputs to marketing differentiation. Figure 10-1 shows how these four activities are interrelated.

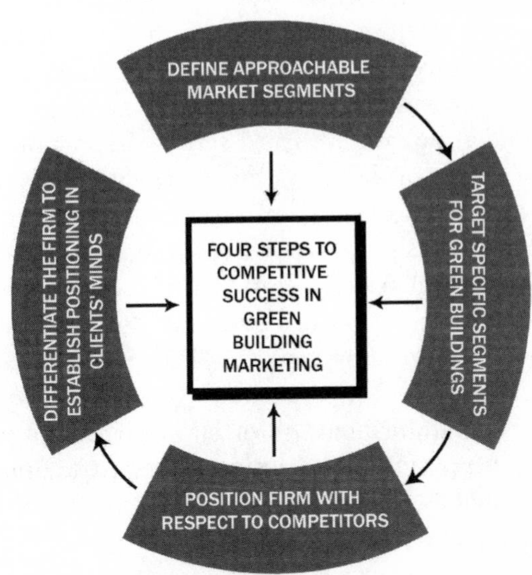

Figure 10-1. Segmentation, Targeting, Positioning and Differentiation

SEGMENTATION

Marketers are always trying to understand and segment markets in order to focus on the most profitable or available segments. Segmentation variables can include considerations of demographics, geographics, "firmographics," psychographics, and similar issues. In de-mographics, the focus is on the social and economic characteristics of buyers (age, income, race/ethnicity, income, etc.); so far there is little evidence that this approach to segmentation is useful for marketing green buildings. (However, one could argue that those states that are more "liberal" politically are likely to contain a higher number of "change agents" who would be in favor of green buildings, so that in fact socioeconomic characteristics of buyers and decision-makers are relevant; our response is that they are contained already in the geographic category.)

In geo-graphics, the focus is on where people are locating and building; as we saw earlier, there is plenty of evidence that green building activity is concentrated in relatively few places in the United States at this time, such as the West Coast, Mid-Atlantic and Northeast states, with other nodes in the large cities of the South and Southwest, as well as the upper Midwest. The number of LEED project registrations by state, measured against the population of the state would be the first place to look. On this basis, and considering 13 states with at least 28 LEED registered projects (roughly the average number of registrations per state at this time), gives the results shown in Table 10-1. The average number of LEED registrations was about 7.4 per million (280 million people and 2069 project registrations), as of September 2005.

Therefore, geographic location is certainly a prime variable to consider in deciding where to market green building services and products.

Firmographics is a newer term, coined for "business to business" marketing, or B2B. The essential distinctions here are the size of the firm or organization (in terms of revenues, number of locations, number of employees, etc.) to which one is marketing; private, public or nonprofit entity; industry type; and other data similar to demographic data. LEED registrations are clearly more prevalent among public entities (44% of the total), institutions (schools and colleges, hospitals, etc.) and nonprofit groups (20%), compared with 26% of the total project registrations for corporate entities. *Project type* could also be considered a type of "firmographics" segmentation and reflects the fact that most clients prefer to hire design and construction firms with prior experience in their type of project, such as school, college, laboratory, commercial office, and so on.

Psychographics refers to segmenting by lifestyle or propensity to take risk or to tolerate ambiguity in potential outcomes of a green building project. In this classification, a marketer would look for risk-taking personality, people acting as industry leaders, innovators (in the "diffusion

Table 10-1. LEED Registrations per State (selected), as of September, 2005

State	LEED Registrations	Population (Millions)	LEED Registrations per Million
Oregon	102	3.6	28.3
Washington	118	6.1	19.3
Massachusetts	76	6.4	11.9
Arizona	65	5.6	10.8
Pennsylvania	119	12.4	9.6
California	344	35.5	9.6
Maryland	52	5.5	9.4
Michigan	82	10.1	8.1
AVERAGE	**2069**	**280**	**7.4**
Georgia	63	8.7	7.2
Illinois	79	12.7	6.2
New York	115	19.2	6.0
New Jersey	48	8.7	5.5
Texas	88	22.1	4.0

of innovation" sense), as early-stage segments in adopting new technology. Most marketers know who the industry leaders are in given segments and often target them with new ideas such as green buildings, knowing that the vast majority of decision-makers want to see experimentation done successfully using someone else's money before they commit or risk their own.

TARGETING

Targeting is the essential task whereby marketers decide to focus on one or a few segments. This is a critical component of setting marketing strategy: one simply must limit the number of competitive targets, in order to focus on those most likely to be successful. In the case of architecture firms, most specialize in one or a handful of client types (public, private, nonprofit) and market segments (e.g., K12 education, museums, libraries, urban offices, historic preservation and adaptive reuse, healthcare, etc.), so the choice of targets is necessarily limited by the firm's experience and the project resumes of key individuals. Many firms aim to take greater "market share" in a given industry or else extend the geographic reach of their success in tackling a certain type of client, but most firms focus on increasing business from current relationships to grow their business. The more design-oriented the client, the easier it is in general for a smaller "high design" firm to extend geographic reach. Many small design firms successfully work on national and even international levels, typically by teaming with a larger local architecture or engineering firm to provide construction documents and construction supervision. For green buildings, architects and builders who have built an early reputation and history of successful projects are often invited to compete for projects far from home, and they are often successful in doing so, particularly by teaming with local firms.

Prime targets for green building marketing at this time share these characteristics: they are *early adopters* of new technology, they may be *potentially significant users of a new approach* (i.e., they control multiple properties); they may be *opinion leaders* (and therefore be able and willing to sway others, both inside the organization and in a larger community of peers); and they *can be reached at low cost* (e.g., already be clients of a firm or customers for a product).

Since few prospects share all of these characteristics, one has to choose a segment to target, based on a consideration of each of these factors, plus some intangibles, which might include existing relationships, stakeholder activity pushing the prospect to choose green buildings, and market forces pushing local entities to keep up with innovative firms (in such green building "hot spots," for example, such as the Portland and Seattle areas).

POSITIONING

Positioning is the third activity of the STP formula. It takes segmentation and targeting analyses and turns them into messages that go out to clients and prospects. The textbook definition of positioning is "the act of designing the firm's marketing offering and image so that they occupy a meaningful and distinct competitive position in the target customers' minds." (Kotler, 1998, 9th ed., p. 295). In other words, positioning is a communications activity that aims at changing the view of a firm in the mind of a target prospect, in such a way as to create a "difference that makes a difference." These differences have to be *important* (in terms of benefit delivered), *distinctive* (something that not every competitor can claim), *superior* (to other ways to get the same benefit), *communicable* (and somehow visible to prospective clients or buyers), *pre-emptive* (not easily copied by competitors), *affordable* (there is little price difference to get this superior benefit) and *profitable* (the company finds it profitable to be in this market segment). Firms that have positioned themselves successfully as green building experts (through publicizing individual efforts as well as project successes) have found that it is possible to maintain their positioning even as more and more firms try to emulate them (see the case study of Mithūn architects in Chapter 9).

Examples would be firms with certified LEED Gold or Platinum projects or those making the annual "Top Ten" list of the AIA Committee on the Environment (http://www.aia.org/cote). Positioning, then, is what a firm does to take "real facts" and position them in the minds of the targeted prospect; positioning deals with creating perception. In marketing green buildings, positioning is an essential component of a firm's communications strategy and needs to reinforce a single powerful message. Because it is a new industry, green buildings offer the positioning strategy of grabbing a new unoccupied position that is valued by clients and prospects. For example, a firm could claim "the most LEED-registered projects" in a given industry or location, or "the most LEED Accredited Professionals," or "the most LEED Gold projects with a certain product or technology" but then would also have to explain why this is a benefit to a client.

Table 10-2 shows some examples of positioning strategies with examples of firms that use them.

Examining this list of potential positioning strategies makes it quite clear that most firms in the design and construction industry have no clear positioning, and therefore have to compete on their experience with particular building types and their fees. As a result, most design firms have trouble making sufficient profits to grow and to attract major talent from the outside. Many construction firms, especially those in "hard bid" public sector environments, have similar issues.

Figure 10-2 shows positioning strategies that might be adopted by various firms in the green building industry. While the chart refers to design firms, product manufacturers and construction firms also need to construct effective positioning maps, in terms of how they want clients to perceive their product and service offerings, using attributes that make a difference in target-market decision-making.

DIFFERENTIATION

This is an approach to marketing strategy that takes the STP variables and focuses them on particular markets. The differentiation approach to marketing strategy was first popularized in the 1980s by Harvard Business School professor Michael Porter[65] and must be coupled with a specific market, geographic or other focus. In the architecture, engineering and construction professional service industries, the main differentiators for sustainable design are highly qualified people, satisfied clients, high levels of LEED attainment, specific industry and project experience, and the ability to deliver green building projects on conventional budgets. A firm usually needs to show high levels of attainment on the key variables to win major new projects in typically highly competitive situations. Case in point: a recent $5 million green public project north of Seattle drew 24 serious proposals!

A highly acclaimed and seminal work, *The Discipline of Market Leaders*, points out that every firm needs to excel in one of *three key areas of differentiation*: customer intimacy, product leadership and operational

Table 10-2. Strategic Positions[64]

Strategic Positions	Representative Firms
The best	Saks Fifth Avenue, Accenture Consulting
The best value	Hyundai, Schwab (as a discount broker)
Lowest cost	Southwest Airlines, Jet Blue, etc.
Innovation, Pioneer	Boeing, Bank of America, Frank Gehry and OMA (architects)
Product focus	Aamco (transmissions)
Target-specific segment	Gerber (baby food)
Product categories	Gatorade, Oracle
Product attributes	Volvo and Michelin (safety), Crest (whitens)
Product line scope (has everything)	Amazon.com; Barnes & Noble
Organizational intangibles	H-P, Kaiser Permanente (healthcare)
Emotional benefits	MTV, Hallmark Cards
Self-expressive benefits	GAP, Mercedes
Experience of using the product or buying it	Nike, Nordstrom
Personality of the brand	Harley-Davidson, Tiffany
Vs. Competitors	AVIS (vs. Hertz), Ford (vs. GM)

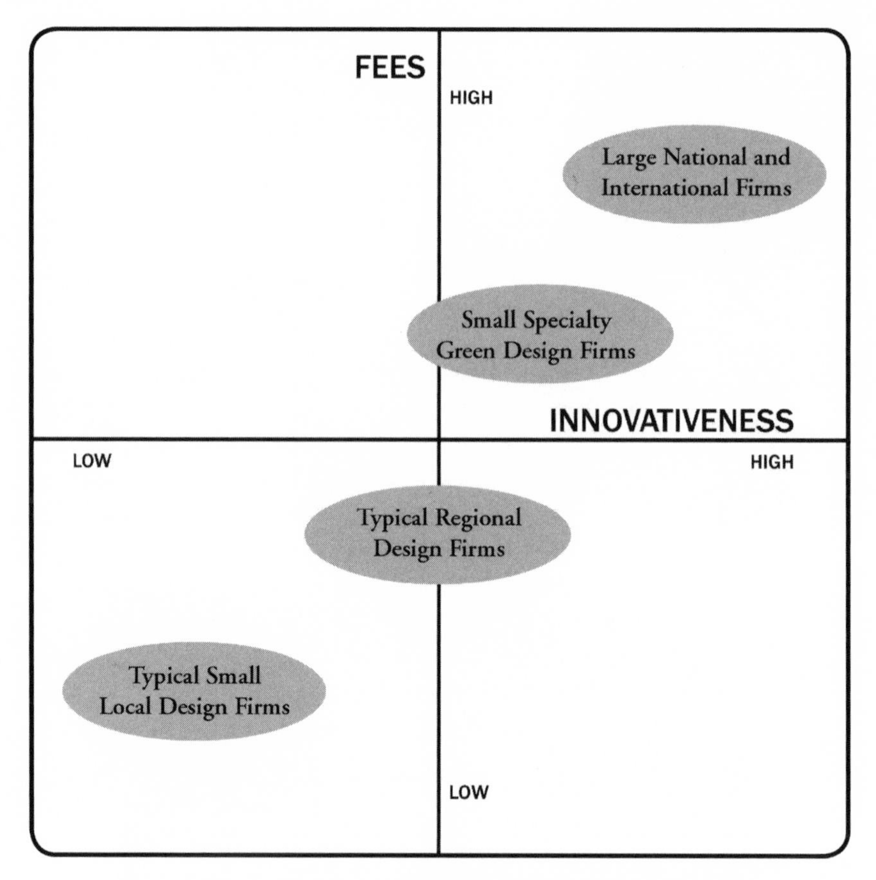

Figure 10-2. Market Positioning Map for Design Firms

excellence, while providing at least good service in the other two areas[66].

- In the area of marketing green buildings, clients expect *intimacy* in the form of established and continuing relationships between clients and architects and builders.

- Firms need to display *operational excellence* in terms of meeting building program goals, budgets and schedules, while achieving specific LEED goals.

- Firms that have a "signature" technological approach can often attract clients who are willing to try new firms who exhibit *product leadership* in the area of sustainable design.

Suzanne Lowe has outlined key differentiation activities for professional service firms in her recent book.[67] Her "top 10" approaches that work for design and consulting firms are:

1. Conducted advertising campaign (to establish/maintain positioning)
2. Added new (to the firm) services that blend into the services of another industry (for example, a consulting engineering firm adding facilities management services)
3. Implemented a formal relationship management program to strengthen the bonds with current clients
4. Merged with another firm, to strengthen the firm's capabilities and reach
5. Managed a public relations campaign (to highlight achievements/reinforce positioning)
6. Extended the firm's services via joint ventures, alliances or referral networks
7. Added new services to the firm within the currently served client base
8. Created a new visual identity (yes, this does work!)
9. Hired specialized, key individuals (often with control of key relationships)
10. Improved or evolved the firm's current services.

Within this list, design and construction firms can find one or more approaches to immediately differentiate their services in the green building industry. Surveys we have done (in Chapter 6) show that the leading firms are particular adept at using differentiation strategies 1, 5, 8 and 9. Improving or evolving the firm's services typically takes place over the course of several green building projects.

SOME THEORETICAL TOOLS

Diffusion Theory

Classical diffusion theory, originally presented in Chapter 3, was developed by Everett Rogers[68] and is widely known among marketers of new technologies. Basically, it posits a group of five distinct personality types who adopt innovations in different ways and at different times. Table 10-3 shows these distinctions. This theory also posits a "normal distribution" of innovation adoption, with a mean time to reach 50% of the potentially available market of typically 10 years.

Table 10-3. Categories of Responses to New Technological Innovations

Name of Category	Percentage of Total	Characteristics
Innovators	2.5	Venturesome
Early adopters	13.5	Respectable
Early majority	34.0	Deliberative
Late majority	34.0	Skeptical
Laggards (or "nevers")	16.0	Traditional

As might be expected, the major issues in determining the rate of adoption of innovation include:

- Relative economic or social advantage (still being debated for green buildings, but generally considered a positive factor)
- compatibility with existing methods (generally this is the case for sustainable design)
- ease of trial at relatively low cost (not the case for new building technologies)

- observability by those who would try it (this is definitely the case for green buildings)
- simplicity of use (which LEED and sustainable design are not, at this time).

Of these five factors, relative economic advantage is the major driver of response to innovation. According to Rogers, there are four overall key factors in determining the rate at which an innovation will spread from the relatively small innovator segment that welcomes new things, to broader segments that are far more risk averse and intolerant of ambiguity.

- The nature of the innovation itself, including its relative advantage
- Communications channels used by subsequent market segments
- Time required for the decision to innovate, the process of adoption to occur and additional adopters to learn about it (the time dimension for completing new buildings, typically two to three years, is short-circuited by the sharing of information from multiple projects, in this case).
- Social system in which the innovation is imbedded, particularly the barriers to innovation.

LEED has gained perhaps 10% of the institutional market for new buildings but scarcely 3% of the corporate market. See also the discussion in the earlier chapters of this book about the state of market adoption of LEED. For the private sector market, the client base can be described as "innovators" and for the public buildings market, the client base is more likely of the "early adopter" category. Even in the public buildings client base, many project managers who supervise large projects could properly be characterized as "late adopters," and will need strong mandates from upper management to pursue sustainable design projects.

The relative advantage of green buildings and LEED has yet to be shown in either of these markets, given the demonstrably higher capital costs and certainly higher certification costs, compared with conventional practice. Certain benefits, such as energy savings, are already a standard part of conventional project "payback" analysis. Benefits appear greater for long-term owner occupants of buildings, but many of the reported and putative benefits are harder-to-measure "soft costs" such as employee productivity, improved morale, reduced absenteeism and illness. From our experience, these benefits have relatively little acceptance at this time among building owners and project financiers.

Anecdotal evidence of benefits is strongly in favor of green buildings, but it has not filtered yet into the general marketplace enough to overcome perceived cost hurdles. Since the green building market is "project based," it may take some time for perceived benefits to find appropriate projects, for a fuller implementation. Oftentimes, adoption of innovation is incomplete, for example, when a technology is acquired (in the way of desired outcomes such as LEED certification) but not deployed into general use; this phenomenon has been called the "acquisition gap" and has been found in a number of technology diffusion studies, which observe that "knowledge barriers impede deployment." This is happening with LEED: 28,000 people have taken the LEED training course, more than 21,000 have passed the LEED AP exam, yet relatively few are actively pushing LEED registration for their design projects, primarily because of their own limited knowledge and fear of client rejection.

In the light of the current state of the market, building owners' and developers' requirements for more *independent cost and performance evaluations of green buildings* are critical for building credibility and overcoming perceived barriers. In my own experience ("Yudelson's Law" for new products), the *expectation of real benefits has to exceed the likelihood of increased costs by 25% or more* to change most decisions in favor of new technologies or methods. Many studies of the psychology of decision-making have shown that consumers and clients are likely to resist change unless the perceived "downside" risk of cost increases and possible performance failures is heavily outweighed by a well perceived "upside" benefit.

Crossing the Chasm

If green building is to enter the mainstream of the "early majority," it must begin to take note of the problems of marketing new technology to the early majority, as outlined by technology marketing guru Geoffrey Moore. It is difficult to go beyond innovators and early adopters to the more general marketplace using the same marketing mechanisms and communications tools as for the smaller, more specialized and risk-tolerant group of innovators and early adopters. The argument here is clearly on the side of simplifying the LEED tool, minimizing annual changes and feature updates, and addressing the risk-aversion of the early majority. A good example might be the update schedule for Microsoft Windows® products, which now appears to be more on a five-year timetable, to avoid upsetting the marketplace.

COMPETITIVE STRATEGY

Most businesses use some variant of the theory of competitive advantage first introduced by Michael Porter of Harvard Business School about 25 years ago. Porter's classic work, *Competitive Strategy* first laid out the three basic building blocks of competitive strategy used by most businesses today. In his work, Porter basically outlines three approaches to winning in the marketplace: differentiation, low cost and focus.

A larger firm can also tie to these three basic strategies a variety of *strategic thrusts,* including *pre-emptive moves* and seeking synergy with other firm activities (such as cross-selling to an existing client a new service or product). The strategic vision's goal is to develop and implement a "Sustainable Competitive Advantage" in the marketplace. Examples of pre-emptive moves would come from larger firms making a major effort to get half or more of their professional staff to become LEED-Accredited Professionals, thereby establishing presumptive expertise in the design of green buildings. (Both Perkins+Will architects, the seventh largest practice in the U.S., www.perkinswill.com, and Keen Engineering (now part of Stantec), one of the largest mechanical engineering firms in Canada, www.keen.ca, have done this). Examples of synergy would include a mechanical engineering firm opening a building commissioning division, an electrical engineering firm specializing in photovoltaic system design, or an architectural firm opening a green building consulting division independent of its regular practice.

In differentiating services, a business seeks to create a difference in the mind of a buyer, with attributes that *make a* difference to that person or organization. For example, we might want to be thought of as the "leading edge" firm or product category; that will limit our market, but sharply define us to buyers who value that attribute, namely the "innovators" of diffusion theory. In today's commercial world, a major task for service firms and for specific technology solutions is to create a *BRAND* that will incorporate those key differences.

Of course, we can create differences for each market segment that we choose to address, since some might value innovation, others low cost, others specific technological choices such as photovoltaics or roof gardens. The author argues that, almost without exception, there are no consumer "brands" in the green building marketplace today. Without a leading brand (and with due apologies to the major companies involved in this business), the average consumer will not want to make a purchase. Even in commercial situations, the lack of a brand can have drawbacks (for example, imagine the confusion in the commercial air conditioning market without major brands such as Trane® and Carrier®).

Low cost of operations gives a firm pricing flexibility. Given the tight budgets of many building projects in the U.S., the ability of design and construction firms and green technologies to compete on price (with low cost) is a valuable asset. These costs may be based on prior project experience, accurate product knowledge, good research, local or state incentives, or a willingness to "pay to get the experience."

The ability to be creative with green building "value engineering" for energy and water savings, along with high levels of indoor air quality, might help an engineering firm to create far more valuable green buildings for the same fee as a more conventionally oriented firm. The ability to specify building-integrated PV systems would fall into the same category, whether for an architectural firm or an engineering firm. (Knowing the costs and the engineering details for PV systems would help an engineering firm to convince owners and architects to move forward with these systems.)

Low-cost advantages might be more sustainable than even branding as a way to compete in the marketplace, but most firms don't have the discipline to operate in this fashion. As a good example of the competitive advantage of lower cost of operations, one can examine the almost unblemished success record of Southwest Airlines. For Southwest, the low prices made possible by lower operating costs have become their primary brand along with "fun." Consider that many of the newer airlines such as Jet Blue, Frontier and Air Tran have even lower costs of operations (expressed as cost per seat-mile) than Southwest, by being very focused in their routes, not trying to be all things to all people, but offering simple air transportation to budget-conscious business and leisure travelers.

Focus is a key competitive strategy, knowing which markets to compete in and which to shun, knowing which clients a firm wants and which it doesn't. Very often, a firm will try to serve too many clients, not really satisfying the clients it really wants by being too unfocused. For most professional service firms (and I have "run the numbers" for my own engineering firm), 80% of revenues come from 20% of the clients served in a given year.

To derive an effective strategy, marketers should consider combining focus with either low cost or differentiation. For example, points of *focused differentiation* can include:

- Regional vs. national firm (many smaller design firms compete nationally by narrowing their focus to one target market, such as museums, libraries, zoos and the like); solar power dealers may certainly compete with a residential vs. commercial focus, or local vs. national. One large commercial PV contracting firm I called in mid-2004 for a quote really impressed me by saying that my job was too small, that they only considered jobs at 100 KW (about $750,000 installation price) or larger. Here is clearly a firm that understands its profitable customer profile and has instructed its salespeople about its decision to serve only larger projects.

- Client types, which can include smaller clients, psychographic profiles (such as early adopter) or those distinguished by strong cultures and values of sustainability. Architects who focus on winning design competitions, for example, clearly seek out adventurous decision-makers for projects that embody a community's or institutions highest aspirations, while others serving the same project types (quite well) do not bother participating in competitions.

- Building or project types (or "vertical markets") such as office buildings, public service facilities (police, fire, jails), secondary education, higher education, health care, labs, cultural centers, retail, hospitality or industrial. Those building types likely to be impacted in the future by far higher peak period electricity rates (up to $0.30 per kilowatt-hour in some of the larger metropolitan areas in the eastern U.S.), such as office buildings and institutional buildings (colleges, public agencies, etc.), might be very good candidates for solar power, particularly in states or utility service areas with significant incentives to offset the higher initial costs of such systems.[69]

- "Signature" green measures, such as photovoltaics, Living Machines or green roofs that a firm commits to bring into play on each project. While it can be dangerous as engineers or architects to "always" bring certain technologies to its projects, *it is more dangerous not to be known for anything in particular*. Branding a firm in the green building arena with specific technology solutions for particular building types and sizes can be an effective marketing measure, allowing such firms to at least make the "short list" for interviews.

- Project size can also be a focus, allowing smaller firms, for example, to compete with larger and more capable competitors. An example might be a focus on operations facilities for public agencies or even green tenant improvements. For smaller projects, many of the larger firms in architecture, engineering and construction are simply uncompetitive in their pricing, since these projects tend to be very "budget challenged."

There is no single competitive response to the growing green building market that is "right" for every firm, as much has to do with the strategic clarity, capability, capital and character of the firm. Nevertheless, *a conscious choice among strategies is vastly preferable to having none*, for that assures only a steady diet of "crumbs" from the table of more decisive firms.

PARTICULAR ISSUES FOR SERVICE MARKETING

Marketing services such as design and construction is inherently different and more difficult than marketing products. Services are unique from products in four ways:

Perishability

Services cannot be inventoried as products can; one lost "man-hour" can never be recovered. Hence firms are always balancing work load with head "count." Given that most service firms have very little capital, it is hard to "staff up" and hope that demand materializes; instead, most firms active in green building design have to carry out a balancing act between having the right people available at the right time, against losing money if demand doesn't materialize to make use of these people's time. This often means that key personnel are assigned too many projects, often in superficial roles.

Inseparability

Services are produced and consumed typically at the same time. In other words, having a firm's associates working on a design at any given time *is* the service purchased by the client. Because of this, "star" performers are often asked for by name by savvy clients and, of course, the star's time is inherently limited. Therefore, *a worthwhile marketing strategy is to make the firm the star*, rather than key individuals. This often requires extensive training to carry out, as well as good internal systems for "technology transfer" from successful projects to new projects. Since the star performers in most professional services firms tend also to be the leading marketers, there is added pressure for them to stay active with projects after they're sold, which is why the design industry is often referred to as a "seller-doer" business, because the seller also has to do the work.

Intangibility

Services are intangible. The quality of a set of green building plans is only discernible to a few, and often not until the building is finished, and all the change orders accounted for. The quality of an integrated design process cannot easily be smelled, tasted, touched or seen, and yet it is critical to the success of projects that have aggressive green goals. To take advantage of this situation, many firms try to create something tangible to point toward the quality of their "intangible" services, such as a quality headquarters building (often involving a LEED-certified renovation or tenant improvement, or even a LEED for Existing Buildings—LEED-EB—project registration); special background studies or signature approaches to projects; marketing communications and marketing collateral materials that consistently emphasize commitment to sustainability; and participation in green building industry associations and events.

Variability

There is no such thing as totally consistent service; good firms have instituted strong quality management programs to try to produce consistent results, but it is always a struggle, because the people in an organization vary greatly in their intelligence, experience, communications skills, personal issues and commitment to client satisfaction. In this respect, hiring, training and retaining the best people is a key marketing strategy for any service organization. See the discussion of the demographics-driven "people problem" in Chapter 19, for an illustration of the critical nature of this problem and how current demographic shifts in the availability of project personnel will affect firms' abilities to deliver green building services in years ahead.

MARKETING STRATEGIES FOR SERVICE FIRMS

Figure 10-3 depicts three interrelated forms of marketing by service firms, including those in the green building industry. Marketing for service firms is very different from marketing for products, because of the amount of client trust and professional competency involved. In the building industry, each "product" is a "one-off" prototype, never to be exactly repeated, whereas in the sale of products, a manufacturer might make a million copies of the same prototype, thereby assuring quality control.

First, service firms carry out "external marketing" to their clients; only this activity is typically considered "marketing" by practitioners in this industry; this type of marketing happens daily when key people at a firm make contact with potential clients. Service firms make a considerable effort to develop marketing collateral materials, place advertising, carry out public relations campaigns, develop client relationship management systems (CRM) and practices, engage in direct mail, newsletters, etc., all to appeal to the client or prospect.

However, the service that is being marketed is actually delivered by individual associates and project teams to the client; this form of marketing can be called "interactive marketing" since the quality of the interaction between client and project team (leading to a "successful" project in the client's mind) is decisive in determining the success of future marketing efforts. A leading academic marketing text puts it this way:

> *Interactive marketing describes the employees' skill in serving the client. Because the client judges service quality not only by its technical quality (e.g., Was the surgery successful?), but also by its functional quality (e.g.,*

Did the surgeon show concern and inspire confidence?), service providers must deliver "high touch" as well as "high tech." (Kotler, 1997, p. 473)

The third aspect to service marketing is "internal marketing," in which the firm trains and indoctrinates its associates in how it expects them to perform for clients, for example, using an *integrated design process* to carry out sustainable design on a given project. This third form of marketing is most often neglected in the architecture, engineering and construction industry. Often, everyone is too busy to invest quality time in training and professional education; however, some firms have made an aggressive commitment to this form of marketing by making sure that most of their professional staff involved with green building projects has passed the test to become a LEED Accredited Professional. In some larger firms, nearly 50% of the professional staff are now LEED APs. Therefore, managing service quality is a key issue in marketing green buildings, whether in the form of design, construction or product marketing.

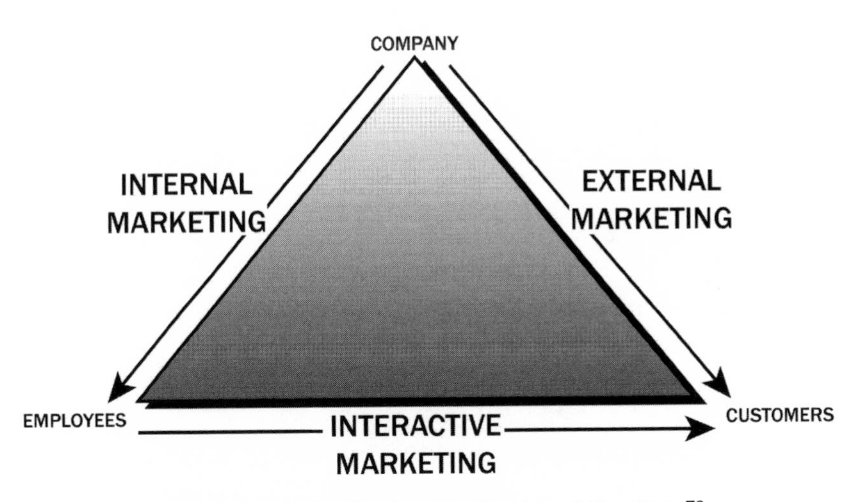

Figure 10-3. Marketing Professional Services[70]

Chapter 11

Selling Green Buildings

How should companies think about marketing and selling high-performance buildings? In all cases, the answer comes down to: who is the buyer? What are their characteristics, unmet needs, motivations and resources? What elements of green buildings do current and potential buyers value most? What are they really buying? How do various customer segments differ in their priorities? What changes are occurring in these priorities? Do the customers for high-performance green buildings fall into any logical groups, based on needs, motivations or characteristics? In Chapter 5, we talked about the "business case" and "value propositions" for green buildings. Let's explore now how these might find their way into the selling of such buildings. (We also take up in Chapter 12 examples of successfully marketed green developments).

MARKET CHARACTERISTICS

The target market for green buildings could anyone who wants to buy or own such a building and has the resources to do so, but we need obviously to find some more specific ways to identify and segment this market. In terms of "the diffusion of innovation," we expect that at this stage of market development, the private sector buyer or owner is going to be an *innovator or early adopter* and somewhat of a risk-taker who is willing to balance the strong case for financial and organizational gain against the risk (and possibly higher costs) of a new approach to building design and construction. Studies that have been done of innovators indicate they tend to be high-status individuals with higher education levels than later-stage adopters. Therefore, *this type of buyer will respond well to a factual presentation of benefits,* will see the longer-term picture, and will likely have done considerable homework of their own before considering the green building approach in a given project.

The institutional or government sector buyer is more likely an *early adopter* of new technology, driven largely by policy considerations, supplemented with the perspective of a long-term owner/occupier/operator of buildings. In other words, the institutional owner is able to look beyond "payback" to the higher value of such buildings and the positive feedback from the stakeholder base: public officials, employees, and the general public. *These people typically are a bit more risk-averse than the "innovator," and tend to rely on social networks for information. They want to see solid cost data and preferably local examples of successful projects.* They will not be the first to act, but because they are not spending their own money, they are willing to take some risks. The nonprofit sector also has an additional motivation: identifying themselves with green buildings has proven to be an effective way to raise money for their building projects and to differentiate themselves in the crowded market for grants and charitable contributions.

Current Market Trends

Chapters 3 and 4 present the current market for LEED/green buildings: more than 2,100 project registrations, representing more than 247 million square feet of buildings, in all 50 states, have registered to use the LEED system, as of the end of September 2005, representing about 3% to 4% of the market for new commercial and institutional buildings in the 2001-2005 time frame. The average LEED-registered building size is therefore about 114,000 sq. ft., with more than one-third less than 80,000 sq. ft. Therefore, there is a significant market for developing relatively smaller high-performance buildings using the LEED green building rating system. Given an educated estimate that perhaps at least the same number of projects uses the LEED system without ever formally registering with the U.S. Green Building Council, it is possible that this market is approaching 4% to 6% of the commercial building new construction starts in 2005. Many of the "unregistered" projects aim to save 20% to 30% of energy use against current codes, putting them squarely in the range of the projected savings from using green building guidelines such as LEED or Advanced Buildings™, especially those that focus on energy savings and indoor air quality improvement.

Looking at LEED-registered buildings using the tools of *market segmentation* yields interesting results. For example, as we saw in Chapter 10, Oregon has the highest per capita LEED registrations in the country, yet it has fairly low energy prices.[71] The next highest state in terms of per capita LEED registrations is Washington, also a state with average to below-average energy prices. (The same comparison holds true when the per capita LEED registrations are compared with state construction activity; see Kats, p. 97). Owing to its large size, California has the highest number of LEED registrations of any state, more than twice as many as any other state.

Clearly, there are places in the United States that have adopted the high-performance building approach; they tend to be in the West (including AZ and CO), the Mid-Atlantic states (MD, DC, VA, PA and NY), and the upper Midwest (IL and MI). For cultural reasons and perhaps owing to higher-than-average utility prices, Massachusetts is also one of the national leaders in LEED registrations. So, *geographic location* is one segmentation variable that does make sense when deciding where to market high-performance buildings. We logically expect that areas with *high utility prices* and strong *local utility incentives* and *state tax benefits* will also be candidates for increased interest in green buildings by tenants, builders and owners, since the "payback" or return on investment (ROI) from energy efficiency investments in new buildings and major renovations would also be higher than in other areas.

Different types of owners and developers are approaching green buildings in their own ways. A great number of public and institutional/nonprofit institutions (including schools and colleges) are using the LEED system or are making significant investments in energy efficiency measures. These owners tend to have taxpayer or donor resources for constructing buildings and have a long-term perspective, as they plan to occupy the buildings they build and therefore reap most of the benefits for themselves.

Other significant specifiers and adopters high-performance office buildings have included large corporations with strong environmental commitments (such as Ford, Toyota, Honda, The Gap) who benefit from the favorable publicity, enhanced employee relations and eventual "payback" of their investment. By contrast, relatively few LEED-registered projects are purely speculative development at a small scale, with the notable exception of some green building "hot spots" such as Portland and Seattle. Most of these projects have to charge market-rate rents and must build green buildings on conventional building budgets. Many developers are figuring out how to do this and how to creatively market energy savings benefits to their tenants. One developer of a renovated 60,000 sq. ft. mixed-use industrial/office space in Portland, Oregon, installed separate electricity and hot water meters for his tenants, so that they could determine how much energy they use and adjust their operations to reduce it. The recent study by the Davis Langdon cost consultants, cited earlier, showed that LEED buildings do not have to carry additional capital cost premiums (although documentation, energy modeling and commissioning costs may still add $2 to $4 per sq. ft. for smaller buildings).[72]

Customer or Client Characteristics

Motivations

What are the market benefits of green buildings and how do these benefits work with the motivations of the various classes of buyers or decision-makers?

- **Securing a Direct Financial Return**. This can take several forms. For example, a public agency could (and often does) view financial return in terms of the long-term cost of ownership, typically using some form of "life-cycle cost" (LCC) analysis, with 4.0% to 5.5% capitalization rates, reflecting today's low cost of public borrowing. A private sector owner such as a large corporation could also be attracted by the Return on Investment (ROI) on energy efficiency investments, using either a corporate weighted average cost of capital or some other criterion such as Internal Rate of Return (IRR), employing a corporate "hurdle rate" for discretionary investments. Green buildings have to compete in a corporation for scarce capital resources and must make convincing cases for the extra investment in financial terms. Other companies use a simpler approach, requiring "payback" of discretionary investments in relatively short time periods of 18 to 36 months. Green building investments for energy efficiency often can provide paybacks of two to four years, with an ROI or IRR exceeding 15% to 25%.

- **Reducing Market Risk**. There may be "risk reduction" benefits to private developers as more and more projects achieve a quicker "lease up" owing to their "green" certifications. For example, the Brewery Blocks project in Portland, Oregon, is a 1.7 (approx.) million sq. ft. "spec" commercial mixed-use project occupying five city blocks just north of the downtown core, built at the site of the former Blitz-Weinhard brewery. The project provides retail, class A office space and residential units, along with more than an acre of much needed underground parking. The project developers completely leased up their flagship commercial office building, and they sold out (nine months ahead of opening) the highest-price condominiums in the city, a $48 million,

15-story building called "The Henry," after old-time brewer Henry Weinhard. (This project was featured in a *USA Today* headline article on March 31, 2004). For the residential units, the developers report that the *energy savings and healthy building features were a factor in the purchase decision for about one-third of the condominium buyers and a determining factor for about 10% of buyers* (Personal communication, Dennis Wilde, Gerding/Edlen Development).

- **Getting a Market Edge**. For speculative developers, having a marketing edge in the form of a green building certification, attesting to the benefits to tenants and users, could prove to be a major financial benefit, in the form of a more presentable leasing proposition. It also protects a seller against charges of "greenwashing," or overstating environmental benefits, and *may* provide some measure of protection against future "misrepresentation" lawsuits. This could be a particular concern to developers, since a significant fraction of high-end condominium buyers tend to be lawyers or friends of lawyers.

- **Enjoying Public Relations Benefits**. Many public agencies and large corporations see public relations benefits from green building certifications. For example, responding to a strong public sentiment for environmental responsibility, the City of Seattle mandated in 2001 that all new public buildings above 5,000 sq. ft. had to achieve at least a LEED Silver certification. The City of Vancouver, British Columbia, passed a similar ordinance in 2004, requiring LEED Gold status. In Portland, the "Earth Advantage" program of Portland General Electric provided strong local public relations benefits for more than 60 commercial projects over a seven-year period, through 2003. In New York City, the "Four Times Square" project by the Durst Organization garnered widespread publicity during the design phase in the late 1990s for its variety of green features and was able to lease up the 48-story office building in 2000 primarily to just two anchor tenants, a large law firm and a major publisher. The Durst Organization now has a similar New York City project, the 2.1 million sq. ft, 52-story, *One Bryant Park,* aiming at a LEED Platinum rating when it is completed in 2008; this time they have been able to enlist Bank of America as a 50% partner and 50% occupier of the office space (see case study in Chapter 12.)

- **Improving Risk Profile**. Many large corporations and most public agencies are self-insured. It makes considerable sense for them to invest in higher levels of indoor air quality, for example, to avoid the potential for future court awards to sufferers of "sick building syndrome." In comparison with buildings from the 1970s, which attempted to increase energy efficiency by lowering indoor air quality through reduced ventilation, today's building engineers know how to achieve high levels of energy efficiency while still exceeding code requirements for ventilation and moisture control. Another form of risk management that relates directly to using green building guidelines is the concern building owners have about future large increases in energy prices, especially during peak summer periods. They are seeing how to meet these concerns with such measures as lower overall energy use, green building controls, "off peak" energy generation from thermal energy storage systems and, in some cases, from on-site generation using combined heat and power (CHP) technologies such as micro-turbines and cogeneration systems.

- **Securing an Indirect Financial Return**. The prospects for increased productivity, reduced absenteeism and reduced employee turnover from high-performance buildings are leading many clients to factor these potential benefits into their decision-making. If one considers that "people costs" are the predominant operating costs in a service economy, often comprising more than 70% of total operating costs for most organizations, then it makes sense to maximize productivity, health and morale with higher-performing buildings, employing such techniques as daylighting, improved lighting levels, greater indoor air quality, operable windows, views to the outdoors, natural ventilation, underfloor air distribution systems, and similar measures. Higher levels of indoor air quality can be marketed to tenants and employees through the certification process for LEED, or through other local or national certification programs such as the Advanced Building guidelines, and through project-specific marketing and communications channels such as brochures, building signage, local news articles and reprints, web sites, etc.

- **"Going with Your Gut."** Many developers are leading the way into high-performance buildings because they feel it's the right thing to do and it's the 'wave of the future.' They hope to create a market advantage, in effect "doing well by doing good." One example is the Hines organization in Houston, Texas (www.hines.com), which builds speculative offices for long-term ownership. Hines has expressed its view in many green building forums that a LEED Silver-certified building will provide a long-term market advantage in terms of lower costs of ownership and a better story to sell to prospective tenants. The buildings owned and managed by Hines' professionals strive to maximize efficiency and minimize energy use in creative and pioneering ways. "Because we're trying to build better quality workplaces for our tenants, we have begun to incorporate more sustainable design features," explains Jerrold Lea, senior vice president, Conceptual Construction, at Hines (personal communication). Based on these accomplishments, in 2003 the company was acknowledged as "Energy Star Partner of the Year" for the third consecutive year by the federal government. By the end of 2004, 89 of Hines' buildings had been recognized by the EPA Energy Star program. In 2004, Hines became the only real estate company to receive the Environmental Protection Agency's Energy Star Sustained Excellence Award. At that time, Hines had five projects in the USGBC's "LEED for Existing Buildings" (LEED-EB) pilot program and five projects in the USGBC's LEED for Core and Shell (LEED-CS) pilot program, two of which had been pre-certified at a Silver level. The rendering of Hines' 1180 Peachtree development in Atlanta shows that green buildings can compete with any type of quality real estate project. Hines is a strong proponent of developers' use of the LEED-CS rating system to give a firm a marketplace edge, even while it acknowledges that there is no extra rent available for green buildings from tenants. Hines believes that building energy-efficient green buildings will help its company-owned real estate to attract better tenants, to have a greater lease renewal rate and to maintain their value for the long run.

UNMET NEEDS

If there are strong customer motivations for considering investments in green buildings, then the marketing task for building developers and facility professionals is to respond to the stakeholders' unmet needs by considering high-performance buildings for their next project. In many

Hines' 1180 Peachtree Development in Atlanta

cases, however, these unmet needs are not well articulated enough to compete with other priorities. Often, it makes sense to use something like the Advanced Building guidelines, the LEED rating system or the *Energy Star* rating system to evaluate a project design, to elevate these concerns to the same level of concern as esthetic or other functional criteria. Often during the course of design and construction, high-performance measures are "value engineered" out of the project owing to cost considerations. Many LEED projects have found, for example, that the client's strong requirement to achieve a certain level of LEED certification has forced the design team toward an "integrated design" approach that places the desired LEED rating at the same level as other budget concerns. This has had the effect of requiring the team to look for cost savings in areas other than energy efficiency, water efficiency or indoor air quality, effectively preserving those investments. Often, early stage design "charrettes" or "visioning" sessions can help to articulate some of the unmet needs by key stakeholder groups.

DIFFERENTIATION AS A MARKETING STRATEGY

In today's marketplace, builders, developers and property owners are all looking for an "edge" in the risky world of development. Often, the edge can be found in the "build to suit" realm, in which developers already have tenants before starting the project and take a fee for their efforts. However, higher returns are often found in the speculative arena, so developers need to think about how to differentiate their product offering from others. Often this is done by offering superior location, amenities, flexibility for expansion and similar approaches. Since these features often can easily be duplicated or matched, high-performance buildings may offer a superior way to differentiate a building product, through the means of a third-party endorsement and/or certification, along with the publicity and marketing benefits of "green buildings."

An example would be the new Genzyme Corporation $140 million, 300,000-sq. ft. research facility, which opened in Cambridge, Massachusetts, in April 2004. This project received a LEED Platinum certification in 2005. The CEO of Genzyme stated that "the business case is that our return comes from the productivity of our people, and the building helps us hire and retain those who can make the right decisions." Because of the interest in green buildings in the Boston area, developer Spaulding & Slye Colliers, has started construction on the $53 million, 8-story *Biosquare Research Building D*, also designed to LEED standards, and claims it is 50% leased already. The project includes 160,000 SF of core & shell for tenant laboratories and offices on an urban medical campus in Boston MA. Robert Dickey, managing director, said "green building is a new market phenomenon, and our interest is to capture the value it creates… They lease better, have less expensive mechanical systems that cost 20 to 40 percent less to operate and, therefore, [give us] a higher effective rent over time.[73]"

High-performance buildings offer a way to create "a difference that makes a difference" in addressing buyer motivations and unmet needs. They offer credibility to buyers who are often very sensitive to "greenwashing" claims by sellers of all types of products and services. A third-party certification or other forms of documentation of the specific green building measures in the project offers "risk reduction" to buyers who may not have the ability or experience to judge a developer's claims for themselves. In addition, the third-party certification provides a broker or agent with some way to back up her claims about the building's benefits for prospective tenants. In the large corporate or institutional setting, the certification offers facilities professionals something to "hang their hats on" when reporting to their upper management about the new green building on which they may be spending extra money.

Green building certifications also create a difference in the minds of local print and broadcast media, which can be used to help "differentiate" a project in the minds of prospective tenants. In this respect, LEED and *Energy Star* have developed recognizable "brands" with which the media have become familiar and which they trust. Such projects are still rare enough in most localities that the media will likely remain interested for another couple of years, giving rise to the possibility of significant news coverage for each project that achieves high levels of certification. Most marketing and public relations professionals know that "free media" is far more credible than advertising and considerably cheaper. News articles also lend themselves to reprints and web site linkages, to create further marketing leverage into the future.

SELLING GREEN BUILDINGS

One thing that developers and facility professionals need is the ability to "sell" their choices to others. Often it is necessary to make a "sales pitch" to people holding the purse strings before embarking on the design and construction of a high-performance building. But as most salespeople know, they have to keep selling even after a contract is signed, or run the risk of "buyer's remorse" after the initial sale. For speculative developers in the world of commercial real estate, it is necessary to use the services of real estate brokers, whose main task is to facilitate transactions for their clients. Brokers need to be equipped with an understanding of the "green" features of the project, why they are important and what benefits they create, so that they will be able to present them to prospective clients or tenants. Brokers specialize in negotiation and communications, so some thought has to be given to integrating the green features into the marketing and sales materials for the building, especially if the developer is trying to recapture some of the investment in energy efficiency with higher rents, for example. Since brokers are not going to become specialists in green buildings, these marketing materials have to be straightforward and readily understandable by those without technical training.

How can this selling best be done? In our view, by making the literature about the features of green buildings fit in with the marketing literature for the project. In some ways, this is uncharted territory, especially in the "spec commercial" building world. Nonetheless, the basic lesson of selling remains: "sell the sizzle, not the steak." For technical features of green buildings, this means spelling out and selling the BENEFITS, rather than the features (see Table 11-1). For example, if a project is saving 40% more energy than a commercial building, then the sale to a CEO or COO is that it's 40% cheaper to operate, has a high return on incremental investment, and offers some protection against future uncertainties in energy prices. If the buyer is a tenant, then the healthier indoor air quality or daylighting needs to be marketed in terms of reduced absenteeism due to illness or disease; if the tenant pays the energy bills, then part of the sale is the reduced total operating cost. This is obviously a "harder" sell, in terms of risk to the promoter of the tenant not valuing the benefit appropriately, so some form of certification is really helpful.

Then marketers need to make use of all the tools available today for sales efforts: a project or building web site, with full explanations of the green features and benefits; email newsletters or "e-zines" regarding the building features, along with links to other sites; streaming video testimonials from the designers and builders (or current tenants); links to favorable newspaper and magazine articles about the project; and where possible, radio and TV coverage. One current apartment project in Seattle features such a web site, along with signage at the construction site, and an informative booklet for prospective renters. The project developers, Harbor Properties and Vulcan (see Vulcan case study in Chapter 12), have secured LEED certification for the project. Table 11-1 shows the selling points for the project, emphasizing location, healthy indoor air, respect for tenants' environmental concerns, and possibly lower utility bills, adapted from the project web site.

CASE STUDY: ALCYONE APARTMENTS, SEATTLE

Opened in the summer of 2004, this 7-story, 162-unit project is selling a "laid back lifestyle in an urban environment" in the proximity of Lake Union, near downtown Seattle. Leasing was completed in about nine months in a soft apartment market, with a very broad age range of tenants. The three main selling points are the units themselves, the quiet location and neighborhood amenities, and the focus on "healthy living." The web site (www.alcyoneapartments.com) features a description of the LEED system and the project's LEED certification. The site also includes a "Sustainability Fact Sheet" that succinctly presents the green features and benefits of the project and is worth showing here. The developers believe that the green features convince people to rent at Alcyone, once they've decided on the location.

ENSURING SATISFACTION POST-SALE

In the institutional setting, the facility manager and design professionals often share the responsibility for occupant satisfaction. Many of the "stakeholders" in a high-performance building (from top executives down to the file clerk) need to know what they are getting in their new building, how it works, what the expected benefits are to them and to their organization, and in some cases, how to make it work. Without a strong pre- and post-occupancy sales effort, it is entirely possible that the benefits of the building will go unrealized and un- or under-appreciated. For example, in a building with operable windows, some thought needs to be given to who will actually operate the windows (or in humid climates, how to tell people in offices when they can or are allowed to open the windows). Some people like it hotter and some like it cooler;

Table 11-1. Sustainability Fact Sheet for the Alcyone Apartments, Seattle

What we (developers) did	How residents will benefit
Commercial quality energy-efficient windows	More temperate spaces with increased natural light, greater thermal comfort, and the potential for lower utility bills
A central gas-fired domestic hot water system	Reduced energy use, which means lower utility bills, vs. individual water heaters in each apartment.
Steel stud metal-gauge framing (instead of wood) to increase durability and eliminate shrinkage	Fewer leaks mean less chance of furniture or fixture damage due to water intrusion and mold contamination (a big issue in a cool, rainy climate such as Seattle)
Low volatile organic compound (VOC) paints and carpets limit the "off gassing" of toxic chemical elements into the indoor air	Improved indoor air quality and a healthier indoor environment
Flexcar (hourly rental) parking spaces, electric vehicle charging stations, and bicycle storage facilities encourage alternative transportation	Reduced demand for parking spaces among tenants and potential cost saving from not having to own or operate a car
A rooftop *Pea Patch* garden with recycled rainwater irrigation provides green space, reduces water consumption, and mitigates stormwater runoff from the roof	A green space amenity with a rooftop garden to grow their own plants
80% of construction waste was recycled, to avoid landfill disposal.	A sustainable living environment from design through construction
Located our project on a convenient urban site utilizing existing infrastructure	The convenience of reduced commute times, public transportation options, and nearby culture and entertainment opportunities

they often work side by side, so some controversy is certainly imaginable. (There is considerable research suggesting that people will often tolerate greater temperature swings from "normal" if they have the ability to control the environment.) In the case of natural ventilation, temperature ranges can often exceed five degrees or more from a "normal" 73F or 74F, and employees need to be prepared to dress cooler in the summer and warmer in the winter. In one LEED Gold project in Portland, Oregon, the *Jean Vollum Natural Capital Center,* the building owner (an environmental nonprofit organization) actually put in a lease provision that the allowable temperature band for the building was 68F to 76F, putting tenants on notice to dress for the season. (This 70,000 sq. ft. project leased up very quickly in 2001 and stays fully leased).

CASE STUDY: WINDMILL DEVELOPMENT GROUP, CANADA

Windmill Development Group (www.windmilldevelopments.com) has initiated two Canadian residential projects, one a 10-story building (*The Currents*) representing a brownfield site restoration (a former dry cleaner) in Ottawa, Ontario, with 43 residential units above a public theater. For this project, the team is attempting to generate zero greenhouse gas emissions (GHG) and achieve a LEED Gold certification under the new Canadian LEED version. The main designer is *Busby Perkins + Will* of Vancouver, British Columbia.

The other project is a more traditional townhome/condo development, *The Bridges,* in Calgary, Alberta, scheduled for occupancy in the fall of 2005. Key advertised features include "unparalleled indoor air quality... high-efficiency appliances... natural lighting optimization... and projected lower operating costs to residents." In particular, the buildings are expected to reduce energy and water costs by nearly 50% compared with a "normal" code building. The partners have prepared a downloadable, eight-page booklet explaining the green features and design approach, in an attempt to educate and persuade prospective buyers.

Developers Jonathan and Jeff Westeinde, along with partner Joe van Belleghem of Victoria, British Columbia, have been active in the Canadian Green Building Council and in the development of LEED in Canada. The Westeindes joined forces with van Belleghem in the summer of 2003, following completion and leasing of the latter's successful

LEED Gold-certified project in Victoria, *Vancouver Island Technology Park*, which converted an older hospital into a high-tech office building and successfully leased it.

Of his approach, van Belleghem says:

> *By utilizing a TBL (triple bottom line) approach we try to put our best foot forward and then expand on that with stakeholders to see if we can find innovative ways to try new green techniques in the building. The key is to be up-front and go out of our way to get involved in a community. In our Calgary project, we put in two affordable housing units as we wanted to demonstrate the importance of integration versus segregation. We ensured the units have the same finishes and design quality. They use 50 percent less energy and 60 percent less water so they will stay affordable as utility prices increase. We worked closely with the city, which purchased the units from us. These units have not affected our marketing of the high-end housing and we're hopeful that the city's leadership will have a positive impact on future affordable housing initiatives.*[74]

From a marketing perspective, these two Canadian residential developments are making a clear attempt to "break out of the pack" and differentiate themselves through their green building status, reduced environmental impacts, and lower "total cost of ownership" for residents. These projects appear to be market-rate developments in every other respect, but the hope appears to be that an integrated design process will deliver enough savings on system costs to pay for the extra green features. The projects also respond to the much stronger Canadian commitment to reduce Greenhouse Gas Emissions (GHG) to as close to zero as possible (through high levels of energy savings and purchases of off-site green power) and to subsidize developers who achieve that goal. The jury is still out, of course, on both the consumer response and the level of LEED certification that will be achieved, and we won't really know more until the projects are finished.

The Windmill Development Group relies on partnerships with local government to provide subsidies and other favors that many developers value (such as higher FAR—floor-to-area ratio) and on extensive pre-development public relations stressing their environmental commitment, to get the potential buyers interested. Jeff Westeinde is quoted as saying, "the challenge for the development industry is to come up with infrastructure that puts less of a load on our environment." He says the firm will only do brownfield redevelopment and, on non-contaminated sites, to build only green buildings[75].

As a result of these approaches, in 2005 the team won a major design competition in Victoria, British Columbia, against tough odds, to build an extensive new project called *Dockside Green* (www.docksidegreen.ca) that will aim at all-Platinum projects for a major new housing, office and commercial center on Vancouver Island.

Chapter 12

Marketing Green Developments

This chapter deals briefly with marketing green developments, in other words, how developers sell multiple-use commercial green projects and residential building projects. The projects profiled take advantage of the classic "3 L's" of real estate: location, location and location; but they also use green building features and branding approaches to stand out in a crowded, competitive market for residential and commercial real estate.

SINGLE-FAMILY RESIDENTIAL DEVELOPMENTS

There are a variety of single-family residential green building developments here and there around the United States. One of the earliest in the United States was a 240-home (20 attached and 220 detached units) development called *Village Homes*, developed by Michael and Judy Corbett

Civano - Solarbuilt Home, Tucson, AZ

in the late 1970s on 70 acres in Davis, California, a university town about 15 miles from the state capital, Sacramento. Village Homes developed passive solar home designs, bioswale-based stormwater management, narrow streets to lessen urban heat islands, integrated bikepaths and walkways through the development, a community garden and many other amenities. It was not duplicated for nearly 15 years[76].

One of the new variety of specifically green developments (for this discussion, we are excluding "New Urbanist" developments that do not have a specific green building focus) was *Civano*, an 820-acre, 2,800 home development in Tucson, Arizona (www.civano.com), developed beginning in 1996 as a master development and currently served by four separate homebuilders. In 2004, *Sunset Magazine* named Civano the "Best New Community" in the West. The community connects people to each other and their surroundings by creating a pedestrian-friendly layout. They use drought-tolerant landscaping with native desert plants such as Palo Verde and mesquite that reinforces a sense of place. There is efficient use of resources including water conservation through rainwater harvesting and xeriscaping, energy-efficient building techniques and the wide use of solar energy in homes. The homes were designed to be 50% more efficient than the 1995 Tucson building code; a 2001 study concluded that they saved $500 to $800 in energy costs per year. Water use was 65% below the average local home, also from the 2001 study.

Key Marketing Principles

There are a growing number of single-family residential developments that take some aspect of green building into account. The *key marketing principles* are the following:

- **A clear sense of place:** the homes and communities must look like they belong to the geographic area in which they are built. In the desert of Tucson, AZ, homes are built with low-impact landscaping, attention to solar control and water conservation, solar water heating and smaller roads to reduce heating of roadways and the local environment during the long hot spring and summer. In an area such as Portland, Oregon, where heating is a large energy user (vs. the need mainly for cooling in Tucson), better insulation, more thermally efficient windows, and more attention to indoor air quality and daylighting mark the green residence.

- **Attention to detail**: site layout and orientation is often critical to long-term energy savings, so a master developer must include restrictive covenants for individual builders to follow. (This is the case at Battery Park City in New York, where following LEED standards is now required for all future development, following the successful market entry of *The Solaire*, a 27-story apartment building.) Marketing green developments often follows directly from the initial master planning studies and site layouts, all the way through setting development standards and monitoring compliance, through marketing during a five to ten-year development period.

- **Business partners**: often a local utility will partner with a developer, particularly to offer energy-efficient and solar home certifications. This is the case in Civano, with Tucson Electric Power offering its *Guarantee Home* program with up to 35% savings on residential electric bills[77]. This program has 30 builders signed up, representing about 25% of the local new home market. In the Portland and Salem, Oregon area, Portland General Electric offers its *Earth Advantage* program to developers who agree to build a home that will test at 15% more energy- efficient than a similar typical local new home[78]. The home will also use low-VOC products and offer a better ventilation system. This program currently represents more than 20% of the new home market.

- **Third-party certification**: there are multiple certifications available for green homes, with many parties vying to upstage or pre-empt the "LEED for Homes" rating system. In addition to utility programs that typically certify energy performance, there is also the federal EPA *Energy Star* program that applies to appliances as well as home performance. There may also be local and state programs available to developers. WCI Communities in Florida uses a statewide building industry certification program to validate its projects (see case study below).

- **Focus on a target customer**: typically a middle-class customer, often a "Gen X" homeowner (25 to 40 year olds) who wants an affordable home in a community with an environmental message, less traffic, a safer environment and a community center. The 55 million LOHAS consumers (Lifestyles of Health and Sustainability -- The LOHAS consumer study is available from the Natural Marketing Institute {www.nmisolutions.com}; more information on the LOHAS consumer is available from www.lohasjournal.com.), or "Cultural Creatives," are a looming target for residential green building developers.[79]

- **Differentiation through branding**: the residential developer or green building promoter engages in an extensive amount of advertising, public relations, certification with a local utility or other program such as LEED, visual and thematic branding and other methods to differentiate itself to its target customer base. Since people have almost an unlimited number of new home choices in residential development, especially in major metropolitan areas, standing out from the crowd is essential to successful marketing. Differentiation must take into account a number of segmentation variables such as consumers' dynamic attitudes, behaviors, product usage, lifestyles and demographics.

Case Study: WCI Communities, Florida

A great example of residential branding is taking place in Florida through the efforts of WCI Communities (named "America's Best Builder 2004" by the National Association of Home Builders -- NAHB). WCI Communities is publicly traded (www.wcicommunities.com) and has developed the brand "WCIgreen," with the tagline "Educate. Innovate. Conserve." With a well publicized green demonstration home in 2003, the company moved ahead in 2004 to build an entire community with green building measures in Venice, Florida, near Tampa on the west coast of the state. Venetian's green model, *Casa Verde,* debuted on Earth Day, April 22, 2004. In July 2004, The *Venetian Golf & River Club* earned *Green Development Design Standard* certification by the Florida Green Building Coalition. In March 2005, this project received NAHB's annual "Outstanding Green Marketing Award."

Since there was no "LEED for Homes" product at that time, WCI used the Florida Green Building Coalition's *Green Home Standards* as its benchmark to attaining green status. Casa Verde demonstrates features that improve indoor air quality such as an electrostatic filter to remove particulates and ultraviolet light treatment of circulated air to kill mold and mildew. Products made from renewable sources are also featured, including bamboo or cork flooring. Materials such as tile on roofs, concrete block exterior walls and spray foam insulation rate well for energy efficiency and durability; steel studs rate well for being recyclable and for not using trees. WCI also added *Energy Star* appliances, higher SEER-rated air conditioners and tinted windows to conserve energy. The next project will be a "Zero Energy Home" that will further explore on-site energy production through the use of solar PV systems.

WCI began its journey to sustainability in 1999, with a directive from the company's CEO, Al Hoffman, Jr., to begin making its homes more environmentally responsible. The company hired Karen Childress as its first Environmental Stewardship Manager shortly thereafter and began to explore options in adding environmental features to its offering of high-end (average price about $500,000) primary and secondary residences for upper-income consumers, both in single-family and condominium units. Since embarking on this mission, WCI has received considerable national media coverage for its commitment to green building and for the results. With the CEO's support, Childress has been able to work with designers and project managers, to integrate green features into a portion of the current offering of homes. What's interesting, she reports, is that there is a heightened level of internal company competition to design ever-greener homes, to meet the CEO's mandate.

In 2000, the Florida Green Building Coalition (FGBC), a not-for-profit organization, moved forward and developed a set of standards as a benchmark for green homes in Florida. Each of these standards requires a demonstration of environmental stewardship at various stages of home construction. WCI has worked closely with FGBC and was the first builder in Florida to commit to building an entire community of certified green homes. In the absence of a national standard for green homes, the building community stepped up and created a Florida version. *This is an approach that other home building marketers may want to emulate.*

WCI has worked closely with FGBC on several projects including construction of "the greenest home in Florida" at its Evergene community. WCI was also the first builder in Florida to commit to building an entire community of certified green homes at Venetian Golf & River Club. In June, 2004 four model homes at Venetian were awarded green certification for exceeding the requirements for Florida Green Building Coalition (FGBC) certification. Demonstration and use of native plant and water conservation earned certification from the University of Florida's *Florida Yards and Neighborhoods* Program (FYN). The models also received certification from the EPA as Energy Star Homes and from Florida Power and Light's *BuildSmart* energy efficiency program.

WCI also illustrates the power of partnerships with other organizations to create standards for communities and give the marketing of them more of an imprimatur of respectability. In 2001, WCI teamed up with the nonprofit environmental organization Audubon International (AI) which operates in 20 countries around the world. AI worked closely with WCI to develop and implement new practices to enhance the sustainability of many of the existing and planned communities.

As part of WCI's commitment, in building 10 new communities in Florida, from conception to completion, it is following the principles of sustainability as defined by AI. Designers and builders of these projects address water and wildlife conservation, water quality assurance, vegetation preservation, energy efficiency, and environmental education and outreach. (Most of these standards deal with the land development itself and not so much with home design). In October 2003, Audubon International presented the *John James Audubon Environmental Steward Award* to WCI for setting a new standard for building green homes and sustainable residential communities. WCI was the first homebuilder to receive this award.

Marketing Issues

WCI also illustrates the difference between a broad marketing strategy and just a public relations strategy, in terms of green homes. It's not yet clear that the link has been made with the homeowner to explore how the marketing principles enunciated above can be incorporated into the basic home design. Currently coastal Florida is a "seller's market," with so many Baby Boomers retiring and seeking warmer climates. Many pay in cash and so are not influenced by mortgage rates. However, they are very concerned with future utility costs and more importantly, with health issues, so that developments that provide information on health benefits and certification of the healthy features in their homes can help establish themselves as the experienced and responsible choice.

WCI sponsored consumer research with the University of Florida's Energy Extension Service in 2002 and found, for example, that more than 75% of consumers say they would pay more for a green product, with 41% willing to pay up to a 10% cost premium for energy and water saving appliances; 87% would pay more to save energy if they recovered their investment within five years[80]. For example, how important is indoor air quality, compared with energy and water savings? One would guess that the higher-end consumer is much more interested in personal health issues than in broader environmental impacts of energy and water use.

Finally, WCI believes that its stated and well publicized environmental commitment helps to insulate it from charges of "greenwashing." Childress states "by seeking certification from third parties, WCI has raised the bar on its already high environmental commitment in all aspects of community development. By treating the land respectfully and building demonstration green homes, we are teaching what makes a building green" (personal communication, September 2004).

WCI's Childress emphasizes the cycle of education (both internal and external); building partnerships with multiple stakeholders including government, universities and nonprofits; outreach to the public through web sites, demonstration homes, exhibits and the like; and marketing the entire program to employees with newsletters, posters and incentives.

Other Residential Projects

In Denver, the new 27-acre Highlands' Garden Village (HGV), Denver's latest planned "New Urbanist" neighborhood, was developed by the Jonathan Rose Companies LLC (www.rose-network.com) on the site of a former amusement park and botanical garden. HGV is a mixed-use community just 10 minutes' drive from downtown Denver. The development includes single-family homes, townhouses, and apartment units that are available to a variety of incomes; it also contains 150,000 square feet of office and retail space. HGV occupies a previously developed but abandoned site, and creates the opportunity for some residents to walk to work; the site is also transit-linked, with its own bus stop. Moreover, all of HGV's building materials—recycled and recyclable—exceed the standards of Colorado's *Built Green* program (www.builtgreen.org). Concrete from site demolition was reused for roadbeds, the landscaping is drought-tolerant native species, and some of the buildings run on alternative energy sources such as wind-generated electricity. The Village's car-share program provides vehicles fueled by compressed natural gas that can be rented by the half-hour. In March, 2003, HGV received the U.S. Environmental Protection Agency's 2002 *Clean Air Excellence* Award.

In a recent interview, HGV developer Jonathan Rose said, "there are no proven facts, but we have found that when we build green homes, they sell much quicker than the rest of the market, and they sell for higher prices... We include not only the environmental qualities of the building, but also being in the right location, having gardens all around...So we're selling both community and green; you can't disaggregate them."[81]

MULTIFAMILY RESIDENTIAL DEVELOPMENTS

Development of green building condominiums is beginning in larger cities of the U.S. In Portland, Oregon, the developers of *The Brewery Blocks*, a five-block mixed-use development of about 1.7 million sq. ft. in five buildings, have registered all five of the initial development projects for LEED certification. The first residential building, "The Henry" is a 15-story, $50 million high-rise that includes a retail base of approx. 11,000 sq. ft. on the ground floor, with three floors of parking above and 11 stories of 123 luxury condominiums priced from $199,900 to $1,180,000. This project has the highest prices for any local condominiums (in excess of $300 per sq. ft.) and was sold out nine months before completion (see Chapter 11). Initial costs were not higher, even though the buildings are aiming at a LEED Gold certification. As for public relations, the project was featured in a cover story in *USA Today* (March 31, 2004), making "mainstreaming green" one of the ultimate goals of project publicity (www.thehenrycondos.com). The project is expected to save 30% annually in energy and water

The Henry Condos, Portland, OR (*Photo courtesy of the author***)**

costs, or about $91,000 per year (about $700 per unit). The Henry also features sustainable materials, including wheatboard cabinets, natural-fiber carpets, certified-wood floors and low-VOC paints and sealants.[82]

Also in the Northwest, Unico Properties, a large property management and development firm, is beginning to convert the historic, 75,000 sq. ft. 1910 *Cobb Building* in downtown Seattle. To better use the building's features and to preserve the history and beauty of this unique building, Unico will upgrade the building's systems and redevelop it into a high-end, 90-unit apartment community, with a renovated retail level on the first floor, with occupancy in the fall of 2006. The building will pursue a LEED Silver rating.[83]

On the opposite side of the continent, New York City's Battery Park City Authority developed *The Solaire,* a 27-story, 357,000 sq. ft. apartment building that has gotten similar publicity, and which rented its 293 units quickly, at 4% to 5% above local market rates[84]. Developed for the Authority by the Albanese Organization, The Solaire (www.thesolaire.com) features extensive use of solar photovoltaic (PV) units and estimates it will cut overall energy use by 35% and peak-period electricity use by 65%, a major savings in a very high-energy-price city. The Solaire received a LEED Gold rating in 2003 and also a "Top Ten" award from the Committee on the Environment of the American Institute of Architects (www.aiatopten.org). The project features an on-site wastewater treatment system, stormwater catchment to irrigate a rooftop garden on the 19th floor, upgraded residential air filtering and a PV system that supplies 5% of the building's peak electric power demand. Each year, 5,000 gallons of treated wastewater is used for landscape irrigation. The marketing for The Solaire included extensive local publicity around the ground-breaking in 2001 and heavy use of a web site, including a construction webcam

The Solaire, New York, NY
(*Photo courtesy of the author*)

during the development. The project's web site makes extensive mention of the green features, including a focus on healthy indoor air, certainly a major concern in New York City. Tax credits and state grants totaled $3.3 million for this $115 million project, built for a construction cost of $247 per sq. ft. and completed in August 2003[85]. After a personal site tour, however, we have to state that this project has a breath-taking location along the Hudson River, with views across the river to New Jersey and New York Harbor, including the Statue of Liberty, and a riverfront public park adjacent to the building, making it a highly unusual example of green marketing.

Another major New York residential high-rise, The Helena, opened early in 2005. According to Architecture Week, 28 September 2005, "A crisp, subtly articulated new form has risen among the towers of New York. The Helena, a 38-story, 580-unit apartment building designed by FXFOWLE ARCHITECTS, formerly Fox & Fowle Architects, brings elegant design and sustainable technologies to a building type often underserved in both these regards." Expected to receive a LEED Gold rating, The Helena is a residential high-rise building designed on sustainable principles and supported by the building's owners, the Durst Organization and Rose Associates. Located near the Hudson River on West 57th Street, The Helena offers sweeping, riverside and cityscape views. The Durst Organization also developed *Four Times Square*, an 1990s green commercial high-rise in New York City.

COMMERCIAL DEVELOPMENTS

We have previously mentioned the residential portion of The Brewery Blocks in Portland. The commercial portion encompasses two city blocks and has been equally successful, with most buildings well on the road to being fully leased in a very soft office rental environmental (with local Class A downtown vacancy rates in Portland hovering around

The Helena, New York, NY
(*Courtesy FXFowle Architects*)

Figure 12-1. Advertisement for a Green Building Developer

15%). Each of the office towers has a commitment to achieving at least a LEED Silver certification. The developer, Gerding/Edlen Development (www.ge-dev.com) has made their commitment to green buildings and sustainable practices a major point of marketing differentiation as shown by the advertisement here, which the company ran in local and regional trade media. In the ad, Gerding/Edlen calls their approach "responsible, efficient, essential," which might be the best way yet to describe sustainable design to the marketplace.

In another development, the *Atlantic Station* project in Atlanta, Georgia, has committed to LEED certification of its projected 6 million sq. ft. of Class A office space in a total of 12 million sq. ft. of total development[86]. Being built by a joint venture of insurer AIG and local Jacoby Development, Atlantic Station (www.atlanticstation.com) has reclaimed a brownfield site from a former steel mill on the north end of Atlanta's central business district.

Case Study: Vulcan Development, Seattle, WA

In Seattle, Washington, one of the largest private landowners and real-estate developers is Vulcan Inc., owned by Microsoft co-founder Paul Allen. Hamilton Hazelhurst is real estate development manager for Vulcan and an architect. The company subscribes to a "triple bottom line" of economy, ecology and equity for its projects. In 2001, Vulcan commissioned the Urban Environmental Institute to produce a *Resource Guide for Sustainable Development* to guide the company's subsequent development efforts. In a recent interview, Hazelhurst says:[87]

> *We believe many sustainable strategies will in fact distinguish us in our market and make us more competitive. For instance, we believe strategies that conserve energy and reduce water consumption will be attractive to tenants in a competitive triple-net market (where tenants pay for these costs directly as pass-through expenses), or to landlords in gross markets where operating costs are factored into the base rent and their bottom lines can benefit directly from savings. On the other hand, landlords in a triple-net market who pursue these strategies must be convinced that they will get a rent premium, experience an earlier lease-up or achieve sufficient long-term value for their investment.*

Here, Hazelhurst succinctly states the business case for green buildings for developers with long-term perspectives. Vulcan also expects that their growing reputation as a "good guy" developer will help in future permitting efforts. In addition, they believe that the green building features, including very detailed economic analyses of the benefits of green buildings, will help them in making proposals to large companies looking for space. Green building certifications help build credibility into their marketing messages. Marketers need to take advantage of these insights to make their case to building owners for green buildings.

Hazelhurst believes that the LEED-CS (Core and Shell) program will help developers such as Vulcan who do not control tenant improvements in their projects. He believes that LEED-CS is another incentive to help educate his firm's business clientele. He states:

"Part of what you do as a core and shell operator is to suggest choices for

tenants, and it's still a challenge to encourage them to build out their piece in a green manner. But it's a key opportunity to educate end users about green principles, and we're developing guidelines about how they can proceed with that."[88]

In their marketing efforts, Vulcan sells green buildings in three ways:

- Return on investment, in terms of reduced operating costs for energy and water
- Value of productivity improvements and employee satisfaction
- *Value-based* sustainable features that a company can use to express its commitment to its employees and to influence the way in which the company will be perceived.[89]

Seattle Biomedical Research Center: Vulcan has completed two buildings that aimed to become LEED-Silver certified, including a 5-story, 113,000 sq. ft. life sciences laboratory facility, *Seattle Biomedical Research Center*, Silver-certified under the LEED-CS pilot program, and a 160-unit apartment project. (The company is starting on additional LEED projects at press time.) Strategies include water and energy conservation, improved indoor air quality, rainwater retention and re-use on site, reflective roofing materials, low-VOC interior finishes, and efficient building systems expected to reduce energy use by 20% to 30%.

Seattle Biomedical Research Center, Vulcan/Harbor Properties
(Photo courtesy of Vulcan Real Estate Co.)

A new project, set to open in 2006, *2200 Westlake,* just north of Seattle's downtown area, is a 360,000 sq. ft. mixed-use project with a hotel, 60,000 sq. ft. of retail and 260 condominium housing units. The project plans increased daylighting (a must in a cloudy climate like Seattle's), operable windows, green roofs, rainwater re-use, low energy and water consumption and environmentally sensitive building materials. As a commercial developer, Vulcan wants 60% to 70% lease commitments before committing to construction, and they feel the green features help this process along.

Case Study: One Bryant Park, New York City

For the *ultimate* in sustainable commercial development, consider the *One Bryant Park* development in New York City, scheduled for completion in 2008 by the Durst Organization (www.durst.org), the developers of the well known *Four Times Square,* an early green high-rise in that same neighborhood. This 2.1-million sq. ft., $1 billion-construction-cost development aims at achieving the highest LEED Platinum status through an entire suite of green building measures. The building at Sixth Avenue and 42nd Street in Manhattan, is scheduled for completion in 2008. Bank of America will be headquartered locally in this new, 51-story, 945-foot-tall skyscraper, which will be called the *Bank of America Tower;* it will occupy nearly half of the building's square footage and is a co-developer of the project.

Among other things, the new 40,000 sq. ft. base-floorplate skyscraper will feature a fresh-air shaft, an advanced double-wall glass skin, an onsite storm-water treatment facility for recycling 100% of incident rainwater, an on-site 4.5-MW cogeneration electrical plant with associated thermal energy storage system, a gray-water cooling system, underfloor air distribution and waterless urinals. The designers are also considering the viability of having onsite a geothermal well and an anaerobic digester for on-site wastewater processing to generate water for toilet flushing.

The project aims to save $3 million in annual energy bills, to conserve 4 million gallons of water per year, and to filter 95% of all air particulates (vs. 35% for a typical building). The development partner, Bank of America, owns 50% of the project and has "bought into" the green building features, primarily as a productivity and employee benefit program, as it plans to house 5,000 employees in the building. Architects Cook+Fox of New York are leading the charge for sustainable design at the LEED Platinum level for this project.

MARKETING OPPORTUNITIES FOR PROFESSIONALS

This litany of residential green building projects should make it clear that there are multiple marketing opportunities for civil engineers, landscape architects, urban design and planning experts, mechanical and electrical engineers, architects, builders and many other building industry professionals, first to understand the business case for green buildings, second to figure out how to accomplish the developer's and builder's objectives within a limited budget, and third to understand the certification options available to such projects. For example, LEED-NC can be used in new and renovated high-rise condo and apartment construction, LEED-H in single-family and low-rise residential projects, and the evolving LEED-ND standard for planning neighborhood districts, campuses and even entire new communities. (See Chapter 16.)

LEED FOR CORE AND SHELL (LEED-CS)

The LEED-CS Rating system's pilot phase was scheduled to last until the end of 2005. The LEED-CS core committee, of which the author is a member, expects to launch the next version, LEED-CS v.2.0, in the spring of 2006. LEED-CS will follow closely the points and models of LEED-NC v.2.2, for identical credits. This rating system allows developers to "pre-certify" their projects, attract tenants and financing, and then complete the project. The main eligibility requirement is that the developer control less than 50% of the build-out. This approach allows a developer, for example, to lease up to half a building to an anchor tenant, and then offer the rest of the space as leasable to anyone else.

In assessing this potential market, USGBC became aware that the criteria and desired timing of certification for the potential Core and Shell applicant was significantly different than that for the typical LEED for New Construction applicant. Most of the buildings that qualify for consideration under LEED Core and Shell are designed often built speculatively, without a specific tenant commitment. For the Core and Shell developer to gain market advantage from their LEED initiatives, they must be able to use their LEED-CS designation as part of their approach to marketing differentiation.

LEED-CS pre-certification allows a developer to say, "I intend to complete the building with these features and at this level of performance,"

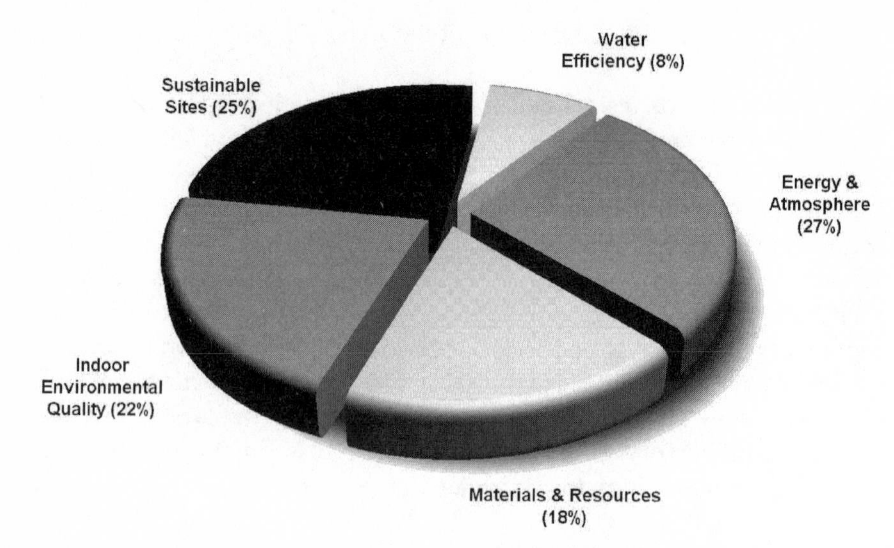

Figure 12-2. LEED CS Core Point Distribution

and the USGBC to say, "If you build the building that you have proposed and document the measures taken, you will be granted a LEED-CS certification at this level." With this distinction, USGBC expects that the developer can more successfully market high-performance sustainable design with a higher level of credibility.

As an example of the benefits of the LEED-CS system, consider the "9th and Stewart Life Sciences Building" in Seattle, Washington. Developed by Touchstone Corporation, this project is an 11-story multi-purpose laboratory and technology building, with 212,000 sq. ft. of laboratory and office space, with parking for 200 cars and 4,000 sq. ft. of ground-floor retail. In spite of the Seattle market's high vacancy rates when this project started programming in 2001, Touchstone was able to move forward by landing a 15-year lease with Corixa Corporation for 65% of the building's rental space.

In terms of environmental performance, the project anticipates a 27% reduction in energy usage, worth about $17,000 in today's energy prices. At a cap rate of 8.5%, this amount of energy savings adds about $200,000, or $1.00 per sq. ft., to the value of the building. A projected 45% decrease in water savings will add an additional $5,500 in annual benefit, adding another $65,000 to building value. The project developer, Douglas Howe of Touchstone, says:

Green building is simply a logical extension of our decision-making process that results in a higher quality and more efficient, high-performance building—one that products a more valuable investment, and enhances the community as well...When the USGBC's LEED program came along, it was a natural progression of our usual practices.

This project also received an award from the National Association of Industrial and Office Properties (NAIOP), a leading national real-estate developers' organization, as the "2004 NAIOP Technology Space of the Year," a testament to its market sense as well as its green attributes. The project was also awarded the "2004 Office Development of the Year" honor from the Society of Industrial and Office Realtors.

9th and Stewart Life Sciences Building, Seattle, WA
(Courtesy Touchstone Corp.)

Chapter 13

Looking to the Future: Sustainable Engineering Design

For the past four decades most architects have designed most buildings in a style known as *Post-Modern,* in which commercial buildings were constructed almost without reference to their environment, facilitated by modern lighting, heating, ventilating and air conditioning systems. These buildings used lots of imported energy, typically in the form of electricity. Buildings were built (and are still being built) without reference to the location of the sun during the day and year, without reference to local wind directions and climate conditions, and without reference to the use of local materials. Sites for new buildings are chosen and buildings oriented with more reference to geometric "urban design" criteria than more natural "ecological design" criteria. As a result, these buildings display a numbing sameness in every geographic region; without names or unusual surface treatments or rooftop displays, one is hard pressed to distinguish them. For many of the inhabitants of these buildings, they are as inhuman a work environment as one could devise, with limited amounts of daylighting, natural ventilation and views to the outdoors.

We all know examples of poorly designed buildings. Two of my favorites:

- A top-floor office for a very wealthy and "important" corporate executive in Portland, with floor to ceiling glass on the 6th floor southwest corner of his headquarters building that requires four separate air-conditioning zones to keep cool in summer (the building has no overhangs or exterior slats to interrupt sunlight penetration.)

- A residential development in Tucson, Arizona, a notoriously hot and sunny climate, with most streets running north-south, so that all homes have a major set of windows facing the western sun, almost guaranteeing that the west-facing rooms will be uninhabitable most afternoons for six months of the year, no matter how powerful the air-conditioning unit. Again, none of these homes in the desert incorporate overhangs or shading of the indoor rooms, though many have a shaded porch on the east side to which the family will have to retreat until the home is habitable in the mid-evening hours!

Meanwhile, where are the engineers in all this?

POST-MODERN ENGINEERING

"Post-modern" engineering has functioned as an "enabler" for the dysfunctional aspects of post-modern architecture. Engineers have used their skills, talents and training to allow architects to build the faceless façades of modern buildings, by ensuring that these buildings would maintain even temperatures 95% or more of the time (± 2F) no matter what the outside temperature, supply only code-mandated amounts of fresh air, provide an overabundance of lighting (100 foot-candles—fc—at one point was the recommended standard from IESNA)—which generated more heat for the buildings' HVAC systems to remove, provision water and power in unlimited amounts, and carry off wastes effortlessly to some receiving treatment plant, body of water or landfill "downstream," to facilitate a mindset that we call "flush without fear."

By doing so, we created a situation in which commercial and residential buildings:

- Use 65% of total U.S. electricity consumption[90]
- Use more than 36% of total U.S. primary energy use[91]
- Produce 30% of total U.S. greenhouse gas emissions[92]
- Generate 136 million tons of construction and demolition waste in the U.S.[93]
- Consume 12% of potable water in the U.S.[94]
- Consume 40% (3 billion tons annually) of raw materials used globally[95]

For further information on the impacts of buildings and the responsibilities of architects and engineers for the global environment, see Edward Mazria's excellent article in *Metropolis*, October 2003, "Turning Down the Global Thermostat: Mazria's Equation.[96]" If we are to develop a truly sustainable society, engineers must relearn their lessons and change the techniques of the past 50 years. What we need today is a new practice: "sustainable engineering."

SUSTAINABLE ENGINEERING DESIGN

In retrospect, what was most unusual about the author's civil engineering curriculum was its focus on the physical sciences and engineering, with no courses taught in biology, ecology, evolutionary biology or any other life sciences. When I did my undergraduate work at Caltech in the 1960s (admittedly a long time ago!), students were all required to take two years of physics and mathematics, one year of chemistry and two years of English and history. Other sciences such as biology and geology were strictly options. Ironically, as a young man growing up in urban Los Angeles, where all the streams were channelized and all the beaches were for surfing and sunbathing, the ideas that bodies of water contained living organisms and a complex ecology was as foreign to me as anything. I can recall only one lone biology-oriented engineering researcher at Caltech who investigated the ecology of marine kelp beds off the coast of southern California and the effect of ocean discharge of treated sewage and sludge on them.

In this first decade of the 21st century, such a curriculum is clearly out of touch with reality, yet the world of engineering practitioners remains mired largely in that era. One cannot get into much trouble by following the ASHRAE *Handbook of Fundamentals* and the building code. A few brave engineers are designing more environmentally sound buildings, but always at the risk of client disapproval and lawsuits if things don't work out.

What would "sustainable engineering" for the built environment look like? (See Table 13-1 at the end of the chapter). It would adopt fundamental principles from physics, chemistry, biology, human factors and psychology, and use them in site development, building design and building operations. As a first effort, engineers would design buildings and their environs to consider:

- Sites would not be located in areas of natural sensitivity, such as habitats for rare or endangered plant or animal species, floodplains, or too near watercourses.

- Buildings would be sited in accordance with passive solar design principles, which could mean a different layout of streets, narrower streets with permeable asphalt to absorb less solar radiation and provide direct recharge of stormwater to aquifers, buildings that are oriented with the long-axis east-west, buildings with less extensive floor plates to allow for more daylight penetration to the interior, buildings with different window treatments on each façade and so on.

- Buildings would exist on available solar income, and ideally produce more power than they consumed.

- Incident rainfall and stormwater would be managed on site whenever possible through a combination of detention, retention and infiltration, to reduce the rate and quantity of off-site flows.

- Water use and sewage generation would be minimized through efficient fixtures; in more adventurous projects, all wastewater would be processed on site.

- Engineers would work with architects to design buildings with daylighting, using efficient controls to mix daylighting with electric lighting to provide adequate lighting levels, typically 30 to 40 fc for most office tasks. (This alone would represent a 20% to 40% reduction in lighting energy use and heat generation from current 50 fc standards)

- Engineers would enable architects to build buildings that used 100% outside air most of the time and that had operable windows. Such buildings would require more elaborate sensors and software, but would "breathe" much more naturally than current buildings with mechanical HVAC equipment.

- Engineers would optimize energy efficiency with longer thresholds for "payback" that would reflect the increased life expectancy of buildings; such thresholds as 20 years for public buildings would allow the inclusion of modest amounts of solar photovoltaic and solar thermal energy systems for most buildings.

- Engineers would design healthier indoor air environments with both structural systems such as underfloor air distribution (raised floor or "access" floor systems) and with mechanical systems providing greater levels of air filtration, separate from thermal control systems.

- Engineering specifications would be more rigorously enforced, with full commissioning of buildings, to ensure that the finished product fully met the design intent.

- Engineers would design outdoor lighting systems that didn't obscure the night sky, by using lower illumination levels and preventing off-site migration of direct-beam illumination.

- Engineers would design performance monitoring and verification systems into buildings, so that the initial design performance could be maintained over a long period of time.

- Engineers would design lighter-weight structures wherever possible, so that fewer materials would be consumed in buildings; where concrete was used for durability or economy, engineers would ensure that as much regionally generated fly ash was incorporated in the concrete as much as possible.

All of these approaches to engineering for the built environment are incorporated or implicit in the LEED performance standards; in that sense, most of them represent "best practices" that most engineers are already familiar with or could learn. Some measures require new tools: for example, a heavy reliance on natural ventilation may require the intensive use of computational fluid dynamics (CFD) modeling, just as evaluating energy efficiency measures requires the use of DOE-2 and other models.

Already, there are some very good examples of what buildings designed to sustainable engineering principles will look like:

- A studio for nearly 150 architects and designers on the Seattle waterfront renovated with no mechanical cooling system at all, just fans for moving air. Admittedly, this is a cool, low-humidity climate, but this firm has a major "green design" orientation and has chosen to "walk the talk" in their new facility, occupied in 2001. (It took them three summers, however, to figure out how to achieve optimum comfort in their new office.)

- A new 275,000 sq. ft. urban high school in the Portland, Oregon, area that incorporates "stack effect" chimneys for natural ventilation and extensive use of daylighting, with a new lighting controller that allows more precise mixing of natural and artificial light.

- Dozens of office buildings in Oregon, Washington, California and British Columbia that use underfloor air distribution systems to conserve energy, allow individual temperature control (a highly prized worker "amenity") of workspaces, and improve indoor air quality by concentrating pollutants above head height.

- A renovated high-rise building in Vancouver, BC, a colder northern coastal climate, with an added double-skin, so that all building windows can open for fresher air, inducing natural ventilation via a "stack effect," without increasing energy use for heating in the winter.

- A new facility designed by William McDonough and Partners for Ford Motor Company at the River Rouge plant in Dearborn, Michigan that incorporates a 13-acre vegetated "green" roof at a cost of $13 million, but is expected to save $47 million in funds that would have to be set aside to control the pollution from the runoff of a "normal" roof.

- A building in Annapolis, Maryland, that reduces water consumption by 90% over a conventional building, through conservation (including a composting toilet) and extensive use of recycled and treated rainwater for sinks (hand washing only) and landscape irrigation. The rainwater storage also doubles as fire protection storage, saving money on a dual system.

So, in what sense will sustainable engineering for the built environment change the way engineers practice? Here are a few examples:

- Engineers will learn new systems that may require different thinking; for example, underfloor air distribution systems operate with far less pressure (0.06″ to 0.10″ w.g., vs. 3.0″ normally) and at higher incoming air temperatures (61F to 63F, vs. 55F normally), reducing fan sizes dramatically.

- Civil engineers will work much more closely with landscape architects and aquatic biologists when stormwater management involves bioswales, green roofs and detention ponds.

- Civil engineers will brush up on their microbiology to handle on-site wastewater treatment systems, either inside the building, or in "engineered wetlands" that simulate natural waste recycling processes.

- Mechanical and electrical engineers will reform their self-image from mere specifiers of mechanical and electrical equipment, to that of "health, comfort and productivity specialists" who are equally adept with natural ventilation and daylighting systems, active and passive solar system design, and filtration against biological contaminants.

- Engineers will become better communicators and psychologists, in order to "sell" their architect and owner clients on the new methods of site and building design and operation, and then to communicate the changes to those who actually manage and operate buildings and properties. They will have to use these skills to convince code officials to let them try new methods of building and site design.

INTEGRATED ENGINEERING

The changes discussed above are clearly significant for most practicing engineers. However, taken together they are still not enough for a truly sustainable society, one that "meets its own legitimate needs while not compromising the ability of future generations to meet their own needs." (United Nations *Bruntland Commission,* 1987). To meet this challenge, adventurous engineers should begin looking into such areas as:

- Designing sustainable communities, in which building lots are laid out for optimum solar orientation, for on-site water and waste management and to minimize automobile use for transportation, perhaps combined with "eco-industrial parks."

- Designing buildings with minimal energy use so that they can exist solely on energy produced from the solar income falling on the site and/or the building.

- Innovative methods of on-site wastewater treatment and rainwater reuse; one advanced biological treatment system in use today, the *Living Machine*™, was developed in the 1970s.

- Looking at building construction and operations as a continuous process, to minimize the use of materials, energy, water and other resources, during the entire lifetime of a building. Current building automation systems have a lot more capability than is currently being used.

- Developing innovative methods of daylighting and natural ventilation for multistory buildings of the kind we continue to build and for those already in existence.

- Finding ways to build earth-sheltered structures that don't produce "dark caves" that most people would rather not work and live in.

- Looking for "Factor 10" opportunities to reduce resource consumption by 90% vs. conventional means, by "thinking out of the box" and re-examining assumptions about what sites and buildings really need.

Principles of ecological design are widely available[97]. These principles force us to go back to basic lessons: all energy should be from solar income; waste is food; water is a resource and not a problem; materials are finite; and human beings need an environment of light, air and connection to the outdoors that should not be compromised in building design. Will sustainable engineering change the way green building engineering is practiced and marketed? Of course it will, and its introduction will require a thorough change in engineering education, training and practice, as well as in marketing and client communications, legal structures and engineering handbooks. This is a race that will be run over the next 10 to 20 years, and those engineers and engineering firms that embrace *sustainable engineering* will be the ultimate winners.

Table 13-1. Comparison of Post-modern vs. Sustainable Engineering

	Post-modern	Sustainable
Buildings	Suburban greenfields New buildings preferred	Urban in-fill Adaptive reuse of building stock
Energy Use	Energy efficiency vs. code Energy efficiency Incremental improvements Utility sources only	Absolute energy use Carbon dioxide production Order of magnitude improvements Distributed (on-site) generation
Economics	First cost is key issue Focus only on the project economics	Life-cycle cost analysis Triple bottom line (economy, environment, social impact)
Ventilation	Forced ventilation Sealed windows High-pressure forced ventilation	Natural ventilation Operable windows Low-pressure displacement ventilation
Climate Control	Components Narrow temperature and humidity fluctuations Consider economics of systems only	Whole systems Wider temperature and humidity fluctuations Look at health and productivity of workers
Environment Materials Selection	Reduced impacts sought No consideration of source, use or disposal Purchase lowest-cost Virgin materials	Restorative systems Life-cycle assessment of sources, use and disposal Purchase from local or regional sources, to promote local economy Use recycled and reclaimed materials, e.g., fly ash in concrete
Water Use	Efficient fixtures	Reclaim and reuse stormwater and graywater
Stormwater	Convey to storm drain for off-site disposal	Detain, retain & recharge on site
Wastewater	Convey to sewer for off-site treatment/ disposal	Treat on-site in constructed wetlands or Living Machines or compact, on-site "bioreactors"

Chapter 14

Marketing Services for LEED-EB Projects

CHANGES IN THE LEED GREEN BUILDING RATING SYSTEM

In this book, we have emphasized the LEED for New Construction version 2.x process (LEED-NC 2.0, 2.1 and 2.2) and guidelines in determining what the markets look like and what they want, including the related LEED for Core and Shell rating system (LEED-CS), which will likely be released in a version 2.0 sometime in mid-2006. However, there are a host of other influences on the horizon for which marketers should start positioning their firms. Some of these are fairly certain, while others are still speculative. In this rapidly changing industry, we can all be sure that change will likely catch us unprepared!

LEED for Existing Buildings (LEED-EB®)

The new LEED-EB standard was in a pilot phase for about two years. In November of 2004, it was "rolled out" as an official standard, LEED-EB version 2.0, following a member balloting process. The primary beneficiaries will be mechanical and electrical engineers (and possibly energy services companies), as LEED-EB focuses heavily on energy use, water use and indoor environmental quality. The LEED-EB standard is focused on the environmental and human impacts of building operating practices, including chemical use, recycling, commuting, purchasing and similar continuing activities of building owners and operators. It is expected to be used most directly by facilities managers to assess the environmental responsibility of their operations and maintenance practices; it may also see significant use by facilities managers and sustainability committees at colleges and universities.

Following the five-category approach of LEED-NC, LEED-EB (July 2005 version) allows for 85 total points (vs. 69 for LEED-NC) and provides a very detailed look at building operations, with considerable focus on energy use, measurement and verification of such use and continuous building commissioning, for example, along with multiple standards relating to green housekeeping and green site maintenance. Figure 14-1 shows the relative point distribution among the five LEED categories for LEED-EB. At the end of this chapter, Table 14-1 shows the various credits. It is unclear at this writing how the LEED-EB standard will evolve; in our judgment right now, it may be over-reaching in terms of the number of environmental issues addressed, making it expensive to implement. We also have doubts about the willingness or financial ability of most organizations to undertake costly reviews of their environmental "footprint" without significant prodding from upper management and without demonstrated organizational and financial benefits. (We have heard, however, that some facilities managers are using LEED-EB to bring more rigor and definition to monitoring their sustainability activities.)

Certification under LEED-EB is intended to allow owners of a large number of buildings, such as the U.S. General Services Administration, schools and colleges, and state general services agencies, to begin certifying their management activities as environmentally sound. It will also allow relatively recent projects that did not certify under LEED for New Construction (LEED-NC) to come back "into the fold."

By October 2004, the LEED-EB pilot project had registered 99 projects,

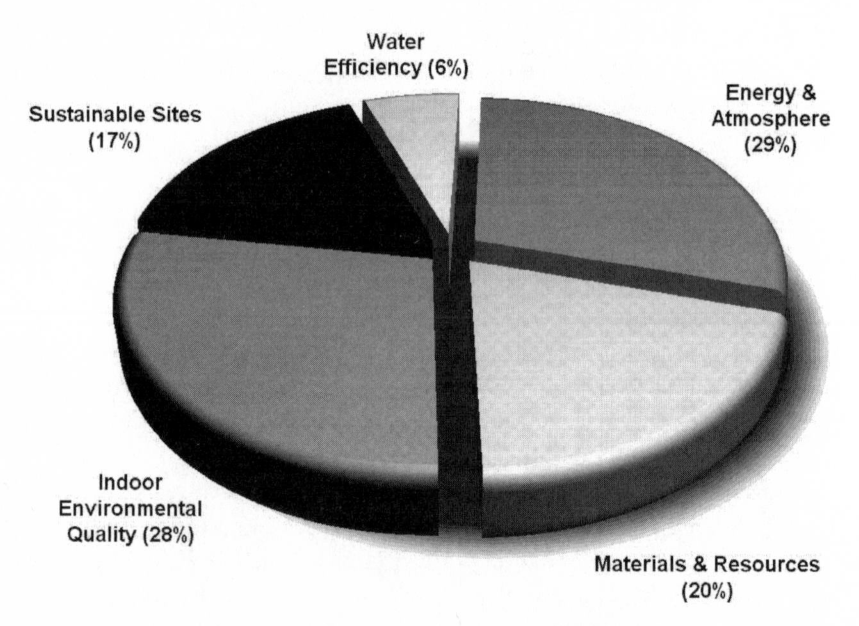

Figure 14-1. LEED-EB v.2.0 Point Distribution

totaling 31.5 million square feet of existing buildings. Pilot buildings were in 28 U.S. states. Five LEED-EB pilot certifications were announced (National Geographic Headquarters in Washington DC; Thomas Properties Group/California EPA Building in Sacramento; JohnsonDiversey Global Headquarters in Sturtevant, Wisconsin; Johnson Controls Brengel Technology Center in Milwaukee; and King County King Street Center in Seattle).

As of the end of July, 2005, LEED-EB version 2.0 had 127 registered projects, 24 certified and more than 38 million sq. ft. of buildings in the program. (USGBC private data dissemination to the author, August 14, 2005).

We foresee a lot of pressure on campuses, for example, for universities and colleges to demonstrate their sustainability commitments by certifying under LEED-EB; we can even foresee when studio classes in architecture, engineering, landscape architecture, construction management and environmental studies will undertake LEED-EB certification for a given building or for campus operations, as part of a year-long course of study. The five basic categories of LEED-EB are modeled after those of the LEED-NC standard, but are modified to focus more on building operations and to remove from evaluation or consideration anything that is "locked in" during original building construction (such as the basic building envelope, glazing, orientation, interior and structural materials and similar features).

In older buildings, for example, as a prerequisite for evaluation, LEED-EB requires upgrading the energy performance to at least the 60th percentile of all buildings of a given type in a specific climate zone; for example, an office building would have to have demonstrated energy use (BTUs/sq. ft., for example) better than 60% of all similar buildings in that region in order to qualify for LEED-EB certification, using the Energy Star calculation methodology. The project would also have to upgrade chillers and other equipment to use CFC-free refrigerants, as another example. Water use would have to be upgraded to new construction standards in order to score any "points" for water efficiency in LEED-EB, which could lead to fixture upgrades and replacements in many cases.

In terms of professional services, commissioning agents, mechanical and electrical engineers, landscape architects and sustainability consultants look to benefit the most. However, the eventual goal of LEED-EB is that all LEED-certified new construction projects will have to be "re-certified" every 5 to 10 years, building in an expanding market of such projects over the next 10 years, as the number of certified projects increases. This may open up opportunities for other professional disciplines to participate.

LEED-EB will also create market demand for certain new products, such as mercury-free (or very low-mercury-content) fluorescent lighting, which are prerequisites under LEED-EB to qualify a building for the certification process. Other requirements in LEED-EB fall in the arena of "Environmentally Preferable Purchasing" and will lead to the purchase of environmentally safe cleaning chemicals, recycled content materials and supplies, and so on. From a marketing perspective, the individuals at architecture and engineering firms who will have to sell the LEED-EB program to building owners and developers will tend to be different than those involved in new construction projects, there will also be a need to certify LEED-EB professionals, to disseminate widely the "Reference Guide" for LEED-EB, and to hold numerous LEED-EB training workshops across the country. In addition, it is not clear at this writing that facility management professionals will want to add another project to their already overflowing plate, given the prevalent lack of adequate budgets in most companies for most routine maintenance and upgrade projects.

Therefore, *we do not look for a significant pickup in activity using the LEED-EB standard until some time in 2006 or even 2007*; even then, market acceptance is not a given, without a significant sales effort by major building controls manufacturers, facility management consultants and building commissioning firms, for example. Unlike LEED-NC, there is no obvious "internal lobby" at a company for spending more money to certify on-going operations, unless an organization has adopted a very strong commitment to sustainable operations and is willing to spend upwards of $50,000 to certify each building, along with making commitments on purchasing policies, landscape policies, cleaning policies, remodeling policies, and similar operational policies that may have company-wide impacts.

Table 14-1. LEED-EB Rating System Categories and Points

Sustainable Sites	Possible Points	14
Prereq 1	Erosion & Sedimentation Control	Required
Prereq 2	Age of Building	Required
Credit 1.1	Plan for Gree Green Site & Building Exterior Management - 4 specific actions	1
Credit 1.2	Plan for Green Site & Building Exterior Management - 8 specific actions	1
Credit 2	High Development Density Building & Area	1
Credit 3.1	Alternative Transportation - Public Transportation Access	1
Credit 3.2	Alternative Transportation - Bicycle Storage & Changing Rooms	1
Credit 3.3	Alternative Transportation - Alternative Fuel Vehicles	1
Credit 3.4	Alternative Transportation - Car Pooling & Telecommuting	1
Credit 4.1	Reduced Site Disturbance - Protect or Restore Open Space (50% of site area)	1
Credit 4.2	Reduced Site Disturbance - Protect or Restore Open Space (75% of site area)	1
Credit 5.1	Stormwater Management - 25% Rate and Quantity Reduction	1
Credit 5.2	Stormwater Management - 50% Rate and Quantity Reduction	1
Credit 6.1	Heat Island Reduction - Non-Roof	1
Credit 6.2	Heat Island Reduction - Roof	1
Credit 7	Light Pollution Reduction	1
Water Efficiency	**Possible Points**	**5**
Prereq 1	Minimum Water Efficiency	Required
Prereq 2	Discharge Water Compliance	Required
Credit 1.1	Water Efficient Landscaping - Reduce Water Use by 50%	1
Credit 1.2	Water Efficient Landscaping - Reduce Water Use by 95%	1
Credit 2	Innovative Wastewater Technologies	1
Credit 3.1	Water Use Reduction - 10% Reduction	1
Credit 3.2	Water Use Reduction - 20% Reduction	1

(Continued)

Table 14-1. LEED-EB Rating System Categories and Points (*Continued*)

Energy & Atmosphere	Possible Points	23
Prereq 1	Existing Building Commissioning	Required
Prereq 2	Minimum Energy Performance - Energy Star 60	Required
Prereq 3	Ozone Protection	Required
Credit 1.1	Optimize Energy Performance - Energy Star 63	1
Credit 1.2	Optimize Energy Performance - Energy Star 67	1
Credit 1.3	Optimize Energy Performance - Energy Star 71	1
Credit 1.4	Optimize Energy Performance - Energy Star 75	1
Credit 1.5	Optimize Energy Performance - Energy Star 79	1
Credit 1.6	Optimize Energy Performance - Energy Star 83	1
Credit 1.7	Optimize Energy Performance - Energy Star 87	1
Credit 1.8	Optimize Energy Performance - Energy Star 91	1
Credit 1.9	Optimize Energy Performance - Energy Star 95	1
Credit 1.10	Optimize Energy Performance - Energy Star 99	1
Credit 2.1	Renewable Energy - On-site 5% / Off-site 25%	1
Credit 2.2	Renewable Energy - On-site 10% / Off-site 50%	1
Credit 2.3	Renewable Energy - On-site 20% / Off-site 75%	1
Credit 2.4	Renewable Energy - On-site 30% / Off-site 100%	1
Credit 3.1	Building Operation & Maintenance - Staff Education	1
Credit 3.2	Building Operation & Maintenance - Building Systems Maintenance	1
Credit 3.3	Building Operation & Maintenance - Building Systems Monitoring	1
Credit 4	Additional Ozone Protection	1
Credit 5.1	Performance Measurement - Enhanced Metering (4 specific actions)	1
Credit 5.2	Performance Measurement - Enhanced Metering (8 specific actions)	1
Credit 5.3	Performance Measurement - Enhanced Metering (12 specific actions)	1
Credit 5.4	Performance Measurement - Emission Reduction Reporting	1
Credit 6	Documenting Sustainable Building Cost Impacts	1

(*Continued*)

Table 14-1. LEED-EB Rating System Categories and Points (*Continued*)

Materials & Resources	Possible Points	16
Prereq 1.1	Source Reduction & Waste Management - Waste Stream Audit	Required
Prereq 1.2	Source Reduction & Waste Management - Storage & Collection	Required
Prereq 2	Toxic Material Source Reduction - Reduced Mercury in Light Bulbs	Required
Credit 1.1	Construction, Demolition & Renovation Waste Management - Recycle 50%	1
Credit 1.2	Construction, Demolition & Renovation Waste Management - Recycle 75%	1
Credit 2.1	Optimize Use of Alternative Materials - 10% of Total Purchases	1
Credit 2.2	Optimize Use of Alternative Materials - 20% of Total Purchases	1
Credit 2.3	Optimize Use of Alternative Materials - 30% of Total Purchases	1
Credit 2.4	Optimize Use of Alternative Materials - 40% of Total Purchases	1
Credit 2.5	Optimize Use of Alternative Materials - 50% of Total Purchases	1
Credit 3.1	Optimize Use of IAQ Compliant Products - 45% of Annual Purchases	1
Credit 3.2	Optimize Use of IAQ Compliant Products - 90% of Annual Purchases	1
Credit 4.1	Sustainable Cleaning Products & Materials - 30% of Annual Purchases	1
Credit 4.2	Sustainable Cleaning Products & Materials - 60% of Annual Purchases	1
Credit 4.3	Sustainable Cleaning Products & Materials - 90% of Annual Purchases	1
Credit 5.1	Occupant Recycling - Recycle 30% of the Total Waste Stream	1
Credit 5.2	Occupant Recycling - Recycle 40% of the Total Waste Stream	1
Credit 5.3	Occupant Recycling - Recycle 50% of the Total Waste Stream	1
Credit 6	Additional Toxic Material Source Reduction - Reduced Mercury in Light Bulbs	1
Indoor Environmental Quality	**Possible Points**	**22**
Prereq 1	Outside Air Introduction & Exhaust Systems	Required
Prereq 2	Environmental Tobacco Smoke (ETS) Control	Required
Prereq 3	Asbestos Removal or Encapsulation	Required
Prereq 4	PCB Removal	Required
Credit 1	Outside Air Delivery Monitoring	1
Credit 2	Increased Ventilation	1
Credit 3	Construction IAQ Management Plan	1
Credit 4.1	Documenting Productivity Impacts - Absenteeism & Healthcare Cost Impacts	1

(*Continued*)

Table 14-1. LEED-EB Rating System Categories and Points (*Continued*)

Credit 4.2	Documenting Productivity Impacts - Other Impacts	1
Credit 5.1	Indoor Chemical & Pollutant Source Control - Reduce Particulates in Air System	1
Credit 5.2	Indoor Chemical & Pollutant Source Control - High Volume Copy/Print/Fax Room	1
Credit 6.1	Controllability of Systems - Lighting	1
Credit 6.2	Controllability of Systems - Temperature & Ventilation	1
Credit 7.1	Thermal Comfort - Compliance	1
Credit 7.2	Thermal Comfort - Permanent Monitoring System	1
Credit 8.1	Daylight & Views - Daylight for 50% of Spaces	1
Credit 8.2	Daylight & Views - Daylight for 75% of Spaces	1
Credit 8.3	Daylight & Views - Views for 40% of Spaces	1
Credit 8.4	Daylight & Views - Views for 80% of Spaces	1
Credit 9	Contemporary IAQ Practice	1
Credit 10.1	Green Cleaning - Entryway Systems	1
Credit 10.2	Green Cleaning - Isolation of Janitorial Closets	1
Credit 10.3	Green Cleaning - Low Environmental Impact Cleaning Policy	1
Credit 10.4	Green Cleaning - Low Environmental Impact Pest Management Policy	1
Credit 10.5	Green Cleaning - Low Environmental Impact Pest Management Policy	1
Credit 10.6	Green Cleaning - Low Environmental Impact Cleaning Equipment Policy	1
Innovation in Operation & Upgrades	**Possible Points**	**5**
Credit 1.1	Innovation in Operation & Upgrades	1
Credit 1.2	Innovation in Operation & Upgrades	1
Credit 1.3	Innovation in Operation & Upgrades	1
Credit 1.4	Innovation in Operation & Upgrades	1
Credit 2	LEED™ Accredited Professional	1
Totals	**Possible Points**	**85**

Certified 32 to 39 points Silver 40 to 47 points Gold 48 to 63 points Platinum 64 or more points

Chapter 15

Marketing Services for LEED-CI Projects

LEED FOR COMMERCIAL INTERIORS (LEED-CI®)

This LEED "product" was released in version 2.0 in November of 2004. Through the first seven months of 2005, 182 projects had registered and 32 had been certified, representing a total of more than 12 million sq. ft. of tenant improvement projects, or about 66,000 sq. ft. per registered project. (USGBC private data dissemination to the author, August 14, 2005).

The effect of LEED-CI will be both on new construction and on building remodels for new tenants who want to meet the higher standard. The USGBC also foresees that developers who certify buildings under the LEED for Core and Shell (LEED-CS) standard will also want to specify that their tenants meet the LEED-CI standard as well. As with LEED-EB, LEED-CI follows the basic five subject format (plus a category for innovation and design process) as the LEED-NC and LEED-CS system, but with fewer credit categories and fewer total points. For example, LEED-CI has only a maximum of 57 points (vs. 69 for LEED-NC) to be attained. There is more focus on furniture and furnishings, lighting and occupancy controls, overall power use of office equipment and lighting, and other factors that might fall under the scope of a typical tenant improvement process. Figure 15-1 shows the different LEED-CI categories.

In terms of professional services, the beneficiaries of LEED-CI are likely to be architects and interior designers first, also mechanical and electrical engineers involved in tenant improvements, as well as the green building consultants who will advise on the referenced sustainability measures and then document the project. As with LEED-NC, there are points available for using significant amounts of certified wood, rapidly renewable materials and recycled or salvaged furniture and furnishings.

In our estimation, *LEED-CI is a very workable standard* and is likely to see considerable use both in tenant improvements in new buildings and in remodels of existing buildings. The demand is likely to be strong

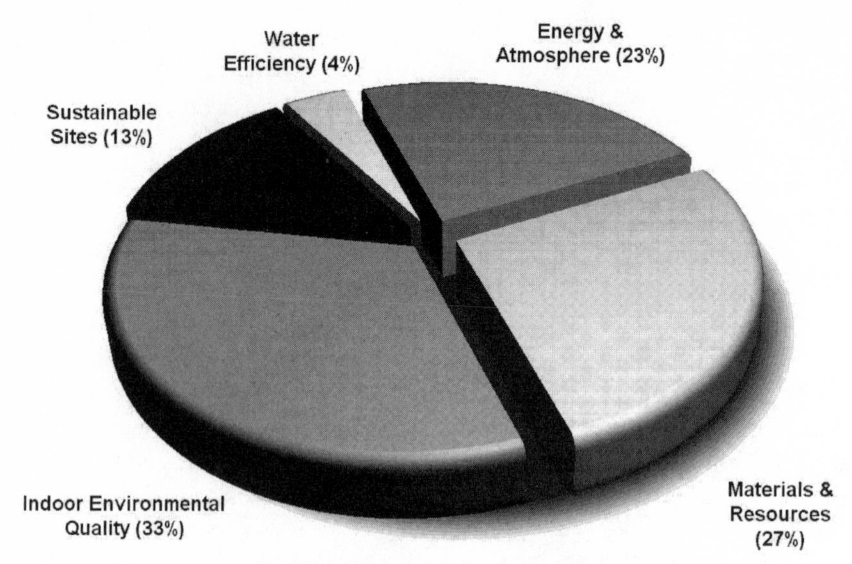

Figure 15-1. LEED-CI Categories—Point Distribution

from corporate users who will see an opportunity to pick up some "sustainability" credits while spending not a lot more than a traditional tenant improvement project would cost. However, since the individuals at architecture and interiors firms who will have to sell the LEED-CI program tend to be different generally than those involved in new construction, there will also be a need to certify LEED-CI professionals, widely disseminate a LEED-CI Reference Guide and to hold specific LEED-CI workshops across the country in this new standard. Therefore, we look for a significant pickup in activity using the LEED-CI standard during 2006 and 2007.

Marketers should move to have their companies certify to LEED-CI standards whenever their company moves its offices; this presents a golden opportunity to secure credit for sustainable design, both inside and outside the firm.

Table 15.1 – LEED-CI Rating System

Sustainable Sites	Possible Points	7
Credit 1	Site Selection - Select a LEED Certified Building - OR -	3
	Locate the tenant space in a building with following characteristics (up to 3 points):	
Option 1A	Brownfield Redevelopment	1/2
Option 1B	Stormwater Management: Rate and Quantity	1/2
Option 1C	Stormwater Management: Treatment	1/2
Option 1D	Heat Island Reduction, Non-Roof	1/2
Option 1E	Heat-Island Reduction, Roof	1/2
Option 1F	Light Pollution Reduction	1/2
Option 1G	Water Efficient Irrigation: Reduce by 50%	1/2
Option 1H	Water Efficient Irrigation: No Potable Use or No Irrigation	1/2
Option 1I	Innovative Wastewater Technologies	1/2
Option 1J	Water Use Reduction: 20% Reduction	1/2
Option 1K	Onsite Renewable Energy	1/2 to 1
Option 1L	Other Quantifiable Environmental Performance	1/2 to 3
Credit 2	Development Density and Community Connectivity	1
Credit 3.1	Alternative Transportation, Public Transportation Access	1
Credit 3.2	Alternative Transportation, Bicycle Storage & Changing Rooms	1
Credit 3.3	Alternative Transportation, Parking Availability	1
Water Efficiency	**Possible Points**	**2**
Credit 1.1	Water Use Reduction - 20% Reduction	1
Credit 1.2	Water Use Reduction - 30% Reduction	1
Energy & Atmosphere	**Possible Points**	**12**
Prereq 1	Fundamental Commissioning	
Prereq 2	Minimum Energy Performance	
Prereq 3	CFC Reduction in HVAC&R Equipment	
Credit 1.1	Optimize Energy Performance - Lighting Power	3
Credit 1.2	Optimize Energy Performance - Lighting Controls	1
Credit 1.3	Optimize Energy Performance - HVAC	2
Credit 1.4	Optimize Energy Performance - Equipment and Appliances	2

(Continued)

Table 15.1 – LEED-CI Rating System

Credit 2	Enhanced Commissioning	1
Credit 3	Energy Use, Measurement & Payment Accountability	2
Credit 4	Green Power	1
Materials & Resources	**Possible Points**	**14**
Prereq 1	Storage and Collection of Recyclables	Required
Credit 1.1	Tenant Space, Long Term Commitment	1
Credit 1.2	Building Reuse, Maintain 40% of Interior Non-Structural Components	1
Credit 1.3	Building Reuse, Maintain 60% of Interior Non-Structural Components	1
Credit 2.1	Construction Waste Management, Divert 50% From Landfill	1
Credit 2.2	Construction Waste Management, Divert 75% From Landfill	1
Credit 3.1	Resource Reuse, 5%	1
Credit 3.2	Resource Reuse, 10%	1
Credit 3.3	Resource Reuse, 30% Furniture and Furnishings	1
Credit 4.1	Recycled Content, 10% (post-consumer + 1/2 pre-consumer)	1
Credit 4.2	Recycled Content, 20% (post-consumer + 1/2 pre-consumer)	1
Credit 5.1	Regional Materials, 20% Manufactured Regionally	1
Credit 5.2	Regional Materials, 10% Extracted and Manufactured Regionally	1
Credit 6	Rapidly Renewable Materials	1
Credit 7	Certified Wood	1
Indoor Environmental Quality	**Possible Points**	**17**
Prereq 1	Minimum IAQ Performance	Required
Prereq 2	Environmental Tobacco Smoke (ETS) Control	Required
Credit 1	Outside Air Delivery Monitoring	1
Credit 2	Increased Ventilation	1
Credit 3.1	Construction IAQ Management Plan, During Construction	1
Credit 3.2	Construction IAQ Management Plan, Before Occupancy	1
Credit 4.1	Low-Emitting Materials, Adhesives and Sealants	1
Credit 4.2	Low-Emitting Materials, Paints and Coatings	1
Credit 4.3	Low-Emitting Materials, Carpet Systems	1

(*Continued*)

Table 15.1 – LEED-CI Rating System

Credit 4.4	Low-Emitting Materials, Composite Wood and Laminate Adhesives	1
Credit 4.5	Low-Emitting Materials, Systems Furniture and Seating	1
Credit 5	Indoor Chemical and Pollutant Source Control	1
Credit 6.1	Controllability of Systems, Lighting	1
Credit 6.2	Controllability of Systems, Temperature and Ventilation	1
Credit 7.1	Thermal Comfort - Compliance	1
Credit 7.2	Thermal Comfort - Monitoring	1
Credit 8.1	Daylight & Views - Daylight 75% of Spaces	1
Credit 8.2	Daylight & Views - Daylight 90% of Spaces	1
Credit 8.3	Daylight & Views - Views for 90% of Seated Spaces	1
Innovation & Design Process	**Possible Points**	5
Credit 1.1	Innovation in Design	1
Credit 1.2	Innovation in Design	1
Credit 1.3	Innovation in Design	1
Credit 1.4	Innovation in Design	1
Credit 2	LEED™ Accredited Professional	1
Totals	**Possible Points**	57

Certified 21 to 26 points Silver 27 to 31 points Gold 32 to 41 points Platinum 42 to 57 points

Chapter 16

Marketing Services for Future LEED Products

LEED FOR HOMES (LEED-H®)

The LEED-H was released as a pilot ("beta") test program in mid-2005. There are certainly "turf conflicts" underway between the National Association of Homebuilders (NAHB), other local and state organizations focused on residential green buildings, the U.S. Department of Energy and the U.S. Green Building Council to develop the "master" residential green building standard. Many homebuilders are already responding to opportunities created by local utility green building rating programs as well as those promulgated with increasing frequency by local governments and state associations (see the WCI Communities case study in Chapter 12). It is not clear that the USGBC will "get to the table" in time to meet the market demand for green building certification, and it is likely that such efforts will remain fragmented for some time to come.

There are currently about 1.8 to 2.1 million new privately-owned housing units built annually in the U.S.[98], with about 1.5 million of those representing single-family detached units, so the market for new LEED registrations is large, assuming LEED could eventually capture 25% of the housing market, its stated goal for all market segments. Even 2.5% of this market (the current approximate level of penetration for LEED-NC) would represent 45,000 to 50,000 new homes certified annually, or about 100 million square feet (given the average size of a new single-family home at 2,000 to 2,500 sq. ft.). The multi-family residential market may turn out to be a greater adopter of the current LEED version 2.x standard, because condominium and apartment developers in many parts of the country are looking for the market edge that LEED can create. They also fit reasonably well with the current LEED-NC standard (which can be applied to residential projects above three stories) and don't have to wait for LEED-H to be developed.

Despite the promise of this market, we do not look for a formal "LEED for Homes" program to emerge from a USGBC pilot test until late 2006 at this point. However, this does NOT mean that there is no interest in green home certifications, but rather that they are likely to emerge under literally dozens of banners around the country for the foreseeable future. Given that the primary USGBC membership is made up of large companies, design firms and construction professionals serving the commercial and institutional markets, it is not clear that the current membership cares sufficiently about a "LEED for Homes" standard to push it forward any faster. Therefore, we believe that most of USGBC's organizational effort the next two years will go toward promoting the new LEED-EB and LEED-CI standards, as well as continuing the refine the LEED version 2.x standard, and that LEED for Homes is likely to lag behind. (This conclusion is not based on any "insider" information, but rather an objective look at the initially slow, then accelerating, development of the LEED-NC program over its first five years).

LEED-NC VERSION 3.0

There are many groups working right now within USGBC on the next generation of LEED, the version 3.0. Realistically, one would expect it to emerge in 2007, similar to the current version but with much more stringent requirements in a variety of sections. (The first attempt to write LEED 3.0 in 2002 was fraught with problems and was sent back to committee by the USGBC Board. There was at that time a tendency to "pile on" one pet issue after another, so as to make LEED a 100-point system, one that would have been rejected, we feel, in the marketplace as overly cumbersome, burdensome and expensive.) Further complications for the next version of LEED may come from the lobbying of various industries to either be included (or not excluded) in the rating system. For example, the vinyl industry fought a proposed credit in the original committee draft for LEED 3.0 that would have granted a point for a "PVC-free" building. Therefore, if we can hazard a guess at this point, again without having any special information, the following changes are likely to appear in LEED version 3.0.

- LEED will become *more stringent* in its requirements, to meet its goal of continuing its *market transformation* orientation; this means that, for example, one point for water efficiency is likely to be awarded at the 30% savings level, vs. 20% today.

- *Relative weightings* of credit categories will likely change a little, with more points likely being awarded to water efficiency and indoor environmental quality.

- The *total number of points* available in the system is unlikely to exceed 75. Compared with today's 69-point system, that change will not represent a huge increase.

- There will be more focus on *absolute levels of energy and water use*, vs. today's relative comparisons. For example, energy use will likely gravitate to the *Energy Star* system of comparing a building's energy use against all other similar buildings in a region. This change will clear up certain anomalies in the current system that, for example, make it difficult to get all 10 energy efficiency points with a building that consumes only 40% of the energy of a "code" building (LEED 2.1 standard), but which uses natural ventilation. Similarly, water use will be measured as total gallons/sq. ft. (or kg/sq.m.) of a building, perhaps for different building types, compared with percentage reductions against today's code. The total energy use analysis of a building will include more focus on "plug loads," currently included in the LEED 2.2 calculations, but increasingly important in assessing energy performance, as base building energy use decreases.

- There will be much more emphasis on *life-cycle assessment* of materials used in buildings, including the energy and environmental impacts of producing, distributing, using and disposing of them. These tools are under active development and aim to provide a more comprehensive way to choose the materials used in a building, considering all environmental impacts over the life cycle of a building.

- *Certain credits will be* dropped, those that hardly any project is using; these include Energy and Atmosphere Credit 5, for example, providing for Measurement and Verification Systems, which will be incorporated instead in LEED-EB. The third credit point for 20% renewable energy use may be scaled back to 15% or even 10% of total building energy use, to encourage solar electric use despite the continuing high cost of photovoltaic (PV) system.

- The "bar" for a certified system could be raised from 40% to 50% of the available points, reflecting the low initial cost of created a Certified building and the increasing sophistication of owners and design teams. That would push up the levels for Silver and Gold certifications, from 50% and 60%, respectively, to 60% and 70%.

- *Controversial issues* such as reducing PVC and vinyl use, elimination of chemicals alleged to be PBTs, and the tradeoff of global warming potential vs. ozone depletion potential for refrigerants, are likely to be addressed in the next version, in spite of potential legal and political complications.

- *Certain prerequisites will be dropped* that represent standard practices in most of the country, and don't add anything but headaches to the LEED system: erosion and sedimentation controls; a ban on CFC use in HVAC systems; and requirements for recycling spaces in buildings, basic ventilation performance and smoking ban in buildings.

- LEED 3.0 will likely adopt *an additional commissioning point*, to stress the importance of design-phase commissioning, acceptance testing, performance verification and training of building operators.

- Competing standards for the same credit category are likely to be recognized in the next version of LEED, ranging from certifying "green" power, to indoor air quality, to a large number of industry-specific product certifications, to requirements for third-parties such as Scientific Certification Systems to certify green claims (www.SCScertified.com). Certifying green products that help meet LEED standards is an area that will acquire far more importance in the next LEED version.

- *Ventilation and indoor air quality will likely increase in importance.* The need for credits that deal more adequately with health, comfort and productivity issues in buildings will likely increase over the next two years and be incorporated in LEED. There is considerable technological progress being made at this time in building space conditioning, and the next version of LEED will address these changes with more sophistication and recognition of emerging design practice.

- A number of *current "innovations,"* such as 95% construction waste recycling, will likely become addressable LEED points, as more and more teams demonstrate their feasibility. This is clearly anticipated by the inclusion of four credit points for innovation included in the current LEED v. 2.x standard.

- There may be a move on the part of the U.S. Green Building Council to require certification by professional auditors, rather than leaving documentation to the design team, owner and contractor. This would be similar to the requirements for certification under the ISO 9000 and ISO 14001 standards for quality management and environmental management.

How should marketers be positioning themselves to take advantage of these changes in LEED?

Those at the leading edge of the green building industry are likely already participating in making the changes in LEED about which we are speculating and will be well positioned to capitalize on them if they do occur. USGBC now allows all members to take part in "corresponding committees," to stay abreast of proposed changes in the LEED system, so savvy companies should make sure that someone on their staff is monitoring the changes that would affect their role in building design and construction.

Firms should be thinking about how to incorporate certain elements of the "new wave" of sustainable design into current projects, without waiting for them to be incorporated into a new LEED standard. This may be a hard sell to a client, concerned only with meeting current LEED requirements, unless it is couched in a larger "sustainability context" and shown to be relevant to stakeholder concerns. An example could be a carbon dioxide emissions mitigation plan (or the purchase of "green tags" for CO_2 offsets) as part of the energy system planning for a new university building, since many colleges and universities are responding at this time to student and faculty concerns about global warming. Finally, companies should be continuing to fund the staff training, in all of the relevant LEED evaluation systems, to ensure that they have qualified people on hand to handle all of the changes likely to occur in green building techniques and strategies in the next five years.

LEED FOR NEIGHBORHOOD DEVELOPMENT (LEED-ND®)

There is considerable interest among urban planners, architects, civil engineers and others for developing a LEED product that addresses larger-scale design issues than assessing just one building at a time. The "LEED for Neighborhood Development" (LEED-ND) committee is quite active, and it is certainly possible that such a certification program will emerge in the 2006-2007 time frame. However, we do not think it is likely to affect the design and construction market much until the 2008-2010 period, simply because very few projects involve entire urban districts; most still include just one building, or perhaps the addition of a building to an established corporate, civic or college campus. What may in fact become certified under LEED-ND are cities' efforts to reduce carbon dioxide production and energy use, for example, and to improve public transportation options, in the form of new urban plans, zoning code changes, water and waste management plans, on-site energy production with central utility plants, and certain infrastructure projects. Once again, it will be necessary for the USGBC to involve established groups such as the Urban Land Institute (www.uli.org) and the Congress for a New Urbanism (www.cnu.org) in the development of these standards.

RESTORATIVE AND REGENERATIVE DESIGN

Way out on the "front lines" of integrated design are a handful of architects, designers and planners looking not just to reduce impacts of buildings on the environment to 30% to 50% below current averages, but to reduce them 75% to 90% or more, to eliminate them entirely, or ideally to begin restoring or healing the environment to "pre-development" conditions. In this context, see the work of New Mexico architect and educator Edward Mazria[99].

In 2003 and 2004, the author participated in such a study for an urban district in the Portland, Oregon area; through careful analysis of solar radiation falling on the district, rainfall and runoff, water and waste flows and related issues, the team was able to chart a path to such a regenerative design for a 40-block area near downtown, over the next 50 years. However, the challenges are daunting: buildings must reduce energy use 75% to 90% over current codes (to be able to live on "solar income" alone); similarly, water use levels must be dramatically reduced to live on the 36" (three feet) of annual rainfall in Portland. What became clear in this study was that new financing and institutional means for effecting these technological changes were also required[100].

If one had to hazard a guess, only the most adventurous or "leading edge" firms need to worry about such issues for the next five years, but it would pay to keep a close eye on these developments, as they may spur other breakthroughs in architectural or engineering design. In much the same way that recent notable buildings and design approaches pioneered by such architects as Rem Koolhaas, Frank Gehry, Norman Foster and Thom Mayne demand that architects and engineers become far more computer-literate than they are today, the envelope for building design keeps getting stretched by computer technology. One direct effect on green buildings in the U.S., as we learn more from green buildings in Europe and Asia, is likely to be *more green spaces incorporated into high-rise buildings* and more incorporation of green spaces with on-site waste treatment on campus buildings.

There are a number of analytical tools and sustainability metrics being developed to assist the design teams investigating the potential for regenerative or restorative design[101]. A Portland group has been working with *The Natural Step* environmental impact assessment tool from Sweden for the past six years, with some interesting results for practical design and construction techniques[102].

PRODUCT CERTIFICATION STANDARDS

Independent, third-party certification of environmental achievements is critical when trying to differentiate a business from its competitors. It bolsters a company's credibility by demonstrating a commitment to the truth and transparency. In this era of aversion to *greenwashing,* consumers, businesses, and government/institutional purchasing agents want to reward the truth. Companies will need to aggressively monitor and contribute to the development of various testing methods, certification standards, and independent certification bodies, in order to certify the "green-ness" of their products and services. By 2006, "corporate sustainability reports" and "green product sheets" similar to "MSDS" sheets will likely be found from most leading manufacturers and service companies at trade shows, meetings, conferences and in marketing collateral. Firms should begin monitoring development and prepare by 2006 or 2007 to start using such information in their marketing activities.

BIOMIMETIC INDUSTRIES

Building on the insights of William McDonough and Michael Braungart in their path-breaking 2002 book, *Cradle to Cradle,* we can foresee the development of products and building materials based on the biological reality that "waste equals food."[103] Products can be made out of biodegradable materials, with no long-lived toxic products, able to break down completely in the environment after their useful life. Materials that cannot break down, such as nylon, can be reused indefinitely; carpets can be returned to the manufacturer and remade into carpets again. Architects and builders will begin to specify such products for buildings. Manufacturing processes themselves would have few or no waste products. The development of "eco-industrial" parks may be the first movement that will affect entire buildings and urban districts, in which all waste products from industry would be re-used by other tenants in the same area; waste heat could be used in greenhouse agriculture or to heat nearby homes, for example.

Biomimicry is emerging as a leading tool in green product developing, studying natural systems to see what they can teach us about resource and energy efficiency in accomplishing tasks. The leading authorities on this topic are Janine Benyus and Hawken and Lovins[104]. Some of their ideas that might influence development of sustainable products include[105]:

- Self assembly: designs which grow themselves and transform over time.

- Solar transformations: molecular-sized solar cells using biological rather than physical means to convert sunlight into electricity.

- The power of shape: the nautilus' logarithmic spiral is influencing the design of turbine and fan blades, which can be made 50% more efficient, to save energy in building ventilation; building design for disassembly; color without pigments and cleansing without detergents.

- Materials as systems, for example, in locally adapted building forms, with structure as finish.

- Material upcycling, with waste from one building as food for another, an idea that has been expressed as "eco-industrial systems," and is influencing the design of industrial parks.

- Ecosystems that grow food and improve fertility, while treating waste, such as Living Machines™.

Chapter 17

The Role of the Professional Engineer in Energy Star®

THE BUSINESS CASE FOR ENERGY STAR BUILDINGS

Deciding to design to green building guidelines is always a challenge, when budgets are so tight and schedules are so compressed. The developer or owner needs to have a really clear idea WHY this is so important. Energy Star is a program developed in the early 1990s by the U.S. Environmental Protection Agency (EPA) to set appliance efficiency standards, which was extended to buildings. By 2004, the label had 70 percent name recognition among consumers, making easily one of the federal government's most successful attempts at creating a consumer brand. A recent analysis for EPA showed that Energy Star buildings save $0.50 per sq. ft., compared with average-performing buildings, and operate 35% more efficiently. These buildings tend to have higher occupancy and continue to save energy over a four-year measuring period.

Let's recap the "business case" for Energy Star buildings:

- **Reduced operating costs**. Energy Star buildings save on operating costs for energy for years to come; with the price of oil up dramatically at over $50 per barrel, and the prospect of peak period electricity prices increasing steadily, it just makes good sense to design the most energy-efficient building possible. Even with "triple net" leases in which the tenant pays all operating costs, it makes sense to offer tenants buildings with the lowest possible operating cost.

- **Risk management**. Energy Star certification can provide some measure of protection against rising prices for electricity and gas, which seem to have become permanent after sharp increases in 2004 and 2005.

- **Stakeholder relations/occupant satisfaction**. Tenants and employees want to see a demonstrated concern for both their well-being and for that of the planet. Smart developers and owners are beginning to realize how to market these benefits to a discerning and skeptical client and stakeholder base, using the advantages of Energy Star certifications and other forms of documentation, including local utility and industry programs.

- **Environmental stewardship**. Being a "good neighbor" is not just for building users, but for the larger community as well. Developers and owners are beginning to see the marketing and public relations benefits (including branding) of a demonstrated concern for the environment, as evidenced by use of the Energy Star brand, recognized by 70% of consumers, according to recent surveys.

- **Increased building value**. Increased annual energy savings will also create higher building values. Imagine a building that saves $37,500 per year in energy costs versus a comparable building built to "code" (this might represent a savings of $0.50 per year per square foot, for a 75,000 sq. ft. building, for example). At a "capitalization rate" (effective discount rate) of 10%, this would add $375,000 (or $5.00 per sq. ft.) to the value of the building! ($37,500/.10 = $375,000). For a small upfront investment, an owner can reap benefits that typically offer a payback of three years or less, and an internal rate of return exceeding 20%, for what is *nearly a sure bet*: energy costs will continue to rise faster than the general rate of inflation and faster than rents can be raised.

- **More competitive product in the marketplace**. There is a dawning realization among speculative developers that green buildings can be more competitive in certain markets, *if* they can be built pretty much on a conventional budget. Whether for speculative or build-to-suit purposes, green buildings with lower operating costs and better indoor environmental quality should be more attractive to a growing group of corporate, public and individual buyers. Energy savings will not replace known attributes such as price, location and conventional amenities, but such features will increasingly enter into tenants' decisions for leasable space and into buyer's decisions to purchase properties.

In 2004, Energy Star claims to have saved $4.2 billion in operating costs in buildings, eliminated the need for 126 billion kilowatt-hours of electrical power use, provided the equivalent of 26,000 MW (megawatts) of electric power plant capacity and prevented 57 million metric tons (carbon equivalents) of greenhouse gas emissions.[106]

In the buildings sector, Energy Star has evaluated more than 23,000 commercial and institutional buildings for energy performance, including 34% of hospitals, 22% of office buildings, 21% of supermarkets, 13% of schools and 9% of hotels. This corresponds to 19% of the square footage of all commercial buildings, more than 450 million square feet, representing a huge data base of building energy use with which engineers can compare their designs. Of this total, 8% have been labeled, a total of 2,300 buildings, which indicates that have achieved an Energy Star rating of 75 or better. (Interestingly enough, in just the past five years, LEED has registered more than 230 million sq. ft., about half of the Energy Star total, with little overlap.)

COMPONENTS OF A SUCCESSFUL ENERGY MANAGEMENT PROGRAM

Based on the successful practices of ENERGY STAR Partners, EPA has identified the key components for a successful energy management program.

- Make a commitment to higher levels of operating efficiency

- Commit to continuous improvement in energy performance
  - Gain "C-level" commitment in the executive suite
  - Focus on portfolio-wide improvement for all buildings
  - Establish an energy policy and program for the organization

- Assess performance and set goals annually
  - The second step in this strategy is to assess performance and set goals
  - U.S. EPA introduced Energy Star as an *energy performance rating system* to meet the need for an objective rating system for each major building type and each of the many climate zones in the U.S.
  - Energy goals can include percent reduction in energy use, by individual building and for an entire portfolio of buildings

— Goals should include financial performance, including return on investment, increase in Net Operating Income (NOI)

- Create and implement an action plan
  — Set priorities across portfolios; let the high-scoring buildings provide "lessons learned" and candidates for Energy Star labels; in the middle, tune up all buildings to yield O&M savings; invest in the poorest performing buildings, because they offer high potential for improvement and high financial returns.
  — Stage improvements, first by commissioning existing buildings, then with lighting, followed by load reduction, upgrades to fans and motors and finally by HVAC system upgrades.

- Evaluate progress, by comparing actual performance to goals, using the Energy Performance Rating tools in Energy Star.

- Recognize success within an organization, motivating people to continue the program.

- Re-assess organizational goals, document progress and identify the next steps.

- Use Energy Star's Portfolio Manager as an online tool for evaluating and tracking the energy performance of buildings over time and obtaining a rating.
  — Normalize building energy consumption: Weather, hours, occupant density, plug load
  — Benchmark for comparison: Similar buildings in national stock
  — Track energy performance over time: Monitor progress
  — Recognize top performing buildings: Top 25% qualify for ENERGY STAR ratings

So, what exactly does the Energy Star performance rating system do? You input a year's worth of energy data and some key characteristics of your customers' facilities, and the system provides a rating on a scale of 1 to 100. In addition to square footage, the energy performance rating normalizes for weather, hours of operations, occupant density, and plug load. A building's energy performance rating can be compared to similar type and use buildings across the U.S.

Eligible building types for an Energy Star rating currently include:

Figure 17-1. components of a Successful Energy Management Program (Source: US EPA)

- Offices (general, bank branch, courthouse, financial center)
- K-12 schools
- Hospitals
- Hotels
- Supermarkets
- Dormitories
- Medical Offices

While this is not the full spectrum of building types, it does give a start toward marketing a firm's green building expertise by focusing on Energy Star labeling wherever it is applicable.

In 2004, EPA expanded the Energy Star program to new building design, as a pilot program, engaging nine commercial building projects. This new program, which allows engineering and architecture firms to put the label, "Designed to meet the ENERGY STAR" on their drawings, represents a potential way for design firms to differentiate themselves to their clients. This label can be used by participating firms if the estimated energy performance of a building design meets EPA criteria. After one year of maintaining superior energy performance, the completed buildings will be eligible to receive the Energy Star designation.

LEED-EB AND THE ENERGY STAR PORTFOLIO MANAGER

LEED-EB has developed equivalency points with the EPA Energy Star system as shown in the table below.

This makes it easy to shoot for an Energy Star rating for upgrades to existing buildings (75 or greater on the EPA scale) and automatically "pick up" four LEED-EB points at the same time. Because of LEED-EB's requirements for "retro-commissioning," it is often easy to improve a building's energy rating enough through simple adjustments to merit a score of 75 or better, providing a "low cost/no cost" approach to garnering these points.

ONLINE FINANCIAL TOOLS

Energy Star also provides online financial tools to help make the case for upgrades:

- *QuikScope* calculates how energy savings can impact net operating income (NOI) and asset value for the commercial real estate market.
- *Cash Flow Opportunity Calculator* estimates how much new equipment or services can be purchased and financed by the anticipated savings cash flows. It compares the costs/benefits of financing the project now vs. later.

In conclusion, energy engineers and green building marketers should pay attention to the Energy Star program, for its objective method of measuring energy performance and for its level of "brand recognition" with top executives and with the public at large.

EPA Rating	**LEED-EB Points**	**EPA Rating**	**LEED-EB Points**
63	1	83	6
67	2	87	7
71	3	91	8
75	4	95	9
79	5	99	10

Chapter 18

Forecasts of Demand for LEED Projects

Here we provide a medium-term forecast, based on "diffusion of innovations" theory and the first five years of data on LEED project registrations, assuming a potentially available market of 120,000 LEED projects (20 years at 6,000 projects per year), shown in Table 18-1 and Figure 18-1. Note that this forecast differs only slightly from the numbers in Tables 3-2 and 4-5 and reflects different approaches to medium-term forecasting. The method used is the Fisher-Pry model of technological substitution, which yields the "S-shaped" curve predicted by the diffusion of innovation theory. The theory predicts a steady slowing of the rate of growth, but a cumulative total LEED registrations more than five times the 2004 year-end total by the end of 2010. If the average LEED project is $11 million in today's dollars (110,000 sq. ft. @ $100 per sq. ft.), then the LEED building market would be $26.5 billion in 2010, encompassing 265 million sq. ft. of project area. Materials sales of all types to LEED projects (at 45% of total construction cost) would equal $11.9 billion.

Table 18-1. Predicted year-end Cumulative LEED Registrations[107]

Year	Cum. LEED-NC Registrations	Annual LEED-NC Registrations	% Growth of Cumulative LEED Registrations
2002	614	344	127%
2003	1061	447	73%
2004	1774	713	67%
2005	2321	547	31%
2006	3021	700	30%
2007	3863	842	28%
2008	5128	1265	33%
2009	7093	1964	38%
2010	9500	2407	34%

Note that the total number of green building projects might be three to five times these amounts, as the number of projects certifying under other guidelines or using the LEED standards but not registering with LEED might be quite significant. These estimates do not consider the growth of LEED-CI, LEED-EB, LEED-H (LEED for Homes) or other project registrations. These are likely to have their own growth dynamics, based on the economic and other benefits they provide to building owners and developers.

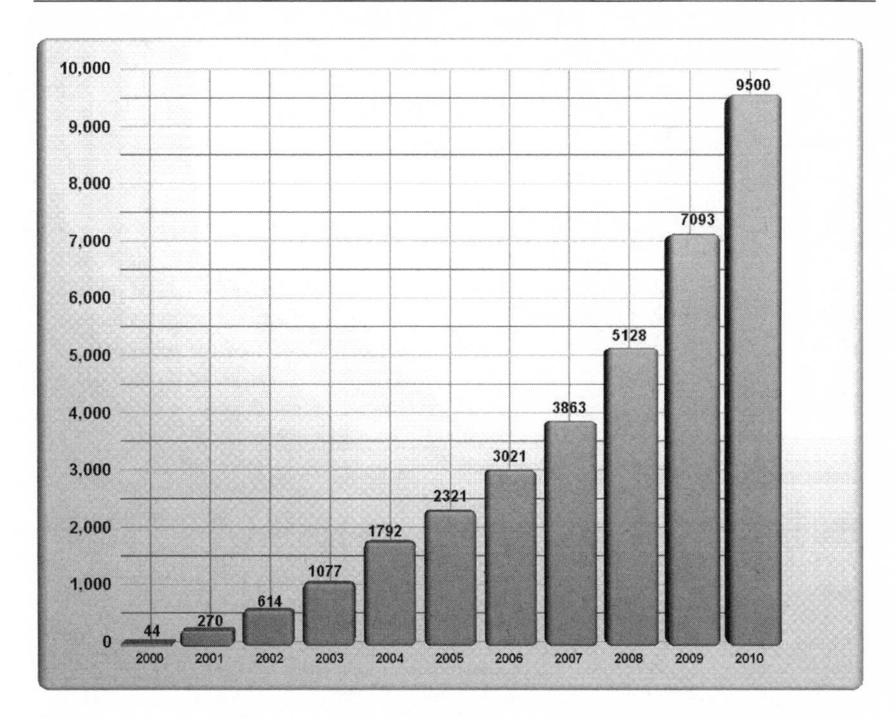

Figure 18-1. Medium-term Projection of Cumulative LEED Registrations

Chapter 19

The People Problem in Marketing Professional Services

No discussion of green building marketing strategies would be complete without a fuller discussion of the "people problem." Without talented people, most firms can't grow and can't take full advantage of their opportunities in this industry. We have heard over and over again that "we have the business, but just can't hire the people."

So, most design firms are experiencing the best of times and the worst of times right now: after a brief recession, business is booming in most market sectors and most parts of the country, but some companies are looking at turning away profitable work because they can't hire enough good people. This situation has been developing over the past 10 years, alongside the growth and then recession of the U.S. economy, but most firms have not responded with a comprehensive strategy to address the "people problem," lately because the design recession of 2002 and 2003 made it easier to hire and then keep designers, and more generally because of a pre-disposition among "baby boomers" who make up design and engineering firm leadership to keep tight control over hiring even in the general expansion of 2004 and 2005, one that looks likely to continue through 2008 or 2009, according to industry sources.

What's going on? Most design firm principals today are "Baby Boomers," those born between 1946 and 1964, making them 40 to 58 years old. For most Baby Boomers, the salient fact of their working lives has been "more people than jobs." Baby Boomers have been competing with their age cohorts for most of their working lives for a relative scarcity of jobs. As a sign of this, real wages in the U.S. did not increase for most of the period from 1973 (when the "Boomers" first began to be a major presence in the work force) until well into the early 1990s, despite the prosperity we associate with the 1980s.

During the late-1990s, the record-breaking U.S. economic expansion created far more jobs than there are people to fill them. One reason for this is the "Generation X" cohort, born between 1965 and 1977, now between 29 and 41 (in 2006). Not only does this group have vastly different expectations for employment, but it is smaller than the Boomer group in absolute numbers. As a result, real wages for this group have started to rise again, for the first time in a generation, and the balance of power between the workforce and firm management will continue to shift dramatically, even though there were layoffs throughout the industry during the 2002-2003 building industry recession. *The main business issue for design and engineering firms is attracting and keeping a high-quality work force; in this sense, a focus on sustainable design is a good way, in the author's experience, to get "short-listed" by qualified candidates.* With the continuing drop in engineering enrollments in U.S. colleges and universities, as well as the lack of foreign-born engineers (most of whom now have better opportunities in their own countries), American engineering firms must do a better job of recruiting, training and retaining key people.

Consider the demographics changes afoot, shown in Table 19-1. By the year 2005, the population in the 25-34 year age group (mostly Gen X), which already fell 9% in absolute numbers from 1995 levels by 2000, will fall another 2.4% by 2005. This is the group of workers on which professional firms depend to "grind out" the daily work. The next age group, 35 to 44 years old, the group which manages most of the work in a professional firm, and which rose from 1995 to 2000 (reflecting the last of the Boomers), will fall 9.4% by 2005 and an additional 8.8% by 2010.

Let's make some sense out of these dry statistics: By 2005, the people available to do the creative daily work of a firm (25 to 34 year olds) falls to a level 11% below that of 1995, 4.6 million fewer people. By 2010, the cadre of people available to manage the daily work of a firm (35 to 44 years old) will fall 14% over 2000 levels, or 6.2 million people.

One might think: OK, we can make up 10% to 15%. But look at what's happening to the economy. Compounded annual growth rates of 3% to 4% make the requirement for workers even greater than today. Consider a 3.5% growth rate, with 2.5% increases in productivity, leaving a need for 1% more workers each year. *By 2010 that requirement means that our worker shortage will be potentially be 24% in the 35 to 44 year old range and 7% to 10% in the 25 to 34 year old range.*

What are the issues for firms? There are three major ramifications:

- Business strategy will have to focus more on profits and less on growth

- Marketing strategy will become hostage to the "people problem"

Table 19-1. Demographic Changes, 1995-2010 (millions)[1]

Year/Age Group	1995	2000	2005	2010
18-24	24.9	26.3 (+5.6%)	28.3 (+7.6%)	30.1 (+6.4%)
25-34	40.9	37.2 (-9.1%)	36.3 (-2.4%)	38.3 (+5.5%)
35-44	42.5	44.7 (+5.2%)	42.2 (-9.4%)	38.5 (-8.8%)

- Human resources will become the most strategic issue for a professional service firm.

Business strategy cannot be predicated solely on adding more people, to grow the firm. More growth will take place via acquisitions and mergers (see for example, the case of Stantec, a Canadian-based public company growing by acquisitions, www.stantec.com.) Businesses will have to continue the late-1990s trend of focusing on key customers and aiming at profitable long-term relationships, with fewer clients and fewer markets covered. The "we do it all" small firm is certainly headed for the "dustbin of history." Business strategy will also rely on outsourcing more and more services; we can even foresee when engineers and designers in less developed nations such as India and China will do the CAD work each evening, after U.S.-based engineers and designers have marked up the drawings; we'll move to the 16-hour and maybe even 24-hour design work day, all enabled by computer technology and the Internet. When outsourcing comes to professional services in a big way, it will change how firms are organized, with today's leaders in the role of account executives rather than designers, more than ever before. (Already, China is a major worldwide source of architectural renderings, doing them at one-third the cost and one-third the time of most American firms).

Marketing professional services is highly dependent on bringing outstanding people to work on a client's problems. With fewer people in the key age ranges, marketing strategy will have to focus less on increasing revenues and more on targeting long-term relationships that have a strong "lifetime value" associated with each client. Marketers will have to become even more involved in creating and selling the image of the firm, since that image will be part of the recruiting effort. (There may be a way to bring systems to bear on design and construction problems in place of people, but these typically take more than a half-decade to develop, test and bring into general practice. One example is the work of some architects to move directly from CAD-generated designs into shop drawings for construction, leaving out the "blueprint" stage of design.) Another, less salutary example is the tendency of building owners and developers to opt for more "design/build" delivery methods as a way to reduce design fees and compress schedules.

Human resources will become elevated as a strategic and management issue. We foresee firms adding an "Executive Vice President, Corporate Development" role that will have command over and responsibility for both marketing and human resources. Every possible means will have to be used to recruit, retrain and retain key people. In our view, these are the "3R's of the New Economy": recruitment, retraining and retention. Keeping and continually retraining a firm's good and average performers is the only viable alternative to constant recruitment.

The good news: if a firm can hold on for the next few years, there is a new generation of people, "Generation Y," that is just as large as the Boomers, and just coming of age. Called by demographers the "Echo" of the Baby Boom, these "Echo Boomers," now under 27, will begin to swell the ranks of younger workers over the next five years, as the numbers of those in the 18 to 24 age group will rise by 14%. However, this group is going to be even more focused on their careers than the "Gen X" group, but paradoxically will demand even more flexibility in scheduling, life style and work style. They are completely Internet-literate and have more information at their fingertips than any of us can imagine.

To summarize: our economy and our professional service firms are facing unprecedented people shortages, and executives must begin to commit significant amounts of management time to preparing design and construction firms to look a lot different five to ten years from now.

The "3R's" of the New Economy will become a mantra for all professional service firms: recruit, retrain and retain our good people.

Appendices

RESOURCES FOR MARKETING DESIGN AND CONSTRUCTION SERVICES

RESOURCES

The number of resources for understanding the green building industry is huge, and the period of self-education is long, so marketers and design professionals need to figure out which resources will keep them on the leading edge. The following is a very brief tabulation of what I use to stay current with this dynamic and fast-growing business.

ORGANIZATIONS

Many green building organizations that provide excellent coverage of this industry.

- *U.S. Green Building Council* (www.usgbc.org) is the largest (5,500 members) and most significant group in the U.S. Publishes the *LEED Reference Guide*, the definitive resources for the LEED system and for green building design in general.
- *Sustainable Buildings Industries Council* (www.psic.org) focuses heavily on schools and residential new construction.
- *Canadian Green Building Council* (www.cagbc.org) covers the same territory for Canada as the U.S. Green Building Council does for the U.S.
- *World Green Building Council* (www.worldgbc.org) leads various country organizations. The *International Initiative for a Sustainable Built Environment* (www.iiSBE.org) is a major international group with a wealth of green building resources.
- The building industry web sites are of course quite valuable, such as the *American Institute of Architects, Committee on the Environment* (www.

aia.org/cote_default), the *American Society of Heating, Refrigeration and Air-Conditioning Engineers*—ASHRAE (www.ashrae.org), and the *Construction Specifications Institute* (www.csinet.org). See also the annual AIA Committee on the Environment "Top Ten" awards, for a sense of the state of the art in green building, www.aiatopten.org/hpb.

- *Collaborative for High Performance Schools* (www.chps.net) has published a fabulous set of design resources in four "Best Practices Manuals" for designing green buildings for schools.

- *BioRegional Development Group* (www.bioregional.com) is working in the UK and also in Portugal on "Zero Energy Developments."

- *New Buildings Institute* (www.newbuildings.org) publishes the *Benefits Guide*, "A Design Professional's Guide to High Performance Office Building Benefits," designed to help architects and engineers talk to their clients about sustainable design for smaller office buildings.

PERIODICALS

Some leading trade magazines and published newsletters include the following (most of these trade publications have on-line newsletters as well). There are numerous magazines covering sustainable design published by professional societies such as AEE, ASCE, ASHRAE, ASLA, IESNA and others.

- *Architectural Record* (http://archrecord.construction.com) is an excellent source for green building information for the mainstream architectural community and a good way for engineers to keep up with the evolving discussion of sustainability among architects.

- *Environmental Design and Construction* (www.edcmag.com), monthly trade magazine.

- *Building Design and Construction* (www.bdcmag.com), monthly trade magazine.

- *Buildings* (www.buildings.com), monthly trade magazine.
- *Consulting-Specifying Engineer* (www.csemag.com), monthly trade magazine.
- *Eco-Structure* (www.eco-structure.com) is a relatively new monthly that also offers excellent covers of green projects and materials.
- *Engineered Systems* (www.esmagazine.com) features "practical applications for innovative HVACR mechanical systems engineers."
- *Environmental Building News* (www.buildinggreen.com), monthly newsletter. Also publishes *GreenSpec*, the most complete guide to specifying green products. Available by subscription for about $200 per year for the entire green building suite of resources.
- *Green Clips* (www.greenclips.com), semi-monthly newsletter, offers a succinct review of five green building stories each issue, and it's free.
- *GreenBiz* (www.greenbiz.com) and its USGBC-affiliated web site, GreenerBuildings (www.greenerbuildings.com), offer excellent industry coverage.
- *HPAC Magazine* (www.hpac.com) covers more technical aspects of heating, plumbing and air conditioning.
- *iGreenbuild* (www.igreenbuild.com), "the voice of sustainable design and construction," is a newer web site with a wide variety of content for green buildings and sustainable design.
- *Metropolis* (www.metropolismag.com), is a monthly design magazine that is expanding its coverage of architecture in general and green building issues in particular.
- *The Sustainable Industries Journal* (www.sijournal.com) provides excellent coverage of green building and sustainable business in the Pacific Northwest.

BOOKS

There are a plethora of good books on various aspects of green buildings and sustainable design. Some of the more significant resources include the following.

- B. Alan Whitson and Jerry Yudelson, *365 Important Questions to Ask About Green Buildings*, 2003 (available from Corporate Realty Design and Management Institute, www.squarefootage.net) deals with the practical questions to ask at each design phase, to avoid precluding consideration of viable green options.

- Jason McLennan, *The Philosophy of Sustainable Design*, 2004, is a good review of the basis for most of today's sustainable design practice. Available from ECOtone Publishing Co., Kansas City, MO, www.ecotonedesign.com.

- Sandra Mendler and William O'Dell, 2000, *The HOK Guidebook to Sustainable Design*, New York: John Wiley & Sons, Inc., the first and one of the more comprehensive guides (412 pages) to the subject.

- U.S. Green Building Council, 2003-2005, *LEED-NC Reference Guides*, version 2.1 and 2.2, edited by Paladino & Associates (Washington, D.C.: U.S. Green Building Council, www.usgbc.org), represents a comprehensive guide to the LEED rating system's current version and an excellent contemporary one-volume resource on sustainable design.

- William McDonough and Michael Braungart, *Cradle to Cradle*, 2002, (New York: North Point Press) details the design philosophy of the "no waste" approach and issues a "manifesto for a new industrial revolution."

- Sim Van der Ryn, *Design for Life: The Architecture of Sim Van der Ryn*, 2005, (Salt Lake City: Gibbs Smith), provides an overview of the present and future of sustainable design from a master practitioner.

WEB SITES AND FORUMS

Perhaps the leading discussion forum today is the Big Green list serve, www.biggreen.org, which features daily postings from a variety of U.S. locations.

- www.greenerbuildings.com, the USGBC and GreenBiz web site, covers the commercial green building industry quite well.
- www.oikos.com, the "Green Building Source," primarily for residential use.
- www.poweryourdesign.com is a web site from New Buildings Institute (www.newbuildings.org) and the source for getting the Advanced Buildings guidelines and the "Benchmark 1.0" tool for designing buildings with energy-efficiency and indoor environmental quality.
- www.wbdg.org is a good source for the U.S. Environmental Protection Agency's "Whole Building Design Guide" and the new 2005 DRAFT *Federal Guide to Green Construction Specs*.
- www.betterbricks.com is an excellent resource for energy-efficient and green building design from the Northwest Energy Efficiency Alliance, www.nwalliance.org, a utility-funded organization that offers hundreds of articles, interviews case studies, and technical resources for sustainable design.

ENDNOTES

Chapter 1

1 Wall Street Journal, October 19, 2005, page B6.
2 U.S. Green Building Council, *Member Update*, September 2005.

Chapter 2

3 www.enr.com/people, accessed October 23, 2005.
4 http://www.stantec.com/news/news_archives/2005/082605.htm, accessed October 2, 2005

5 *Building Design & Construction* Magazine, July 2005, www.bdcmag.com
6 http://www.turnerconstruction.com/corporate/content.asp?d=3747, accessed October 2, 2005
7 See www.cagbc.org.
8 This has led some to claim that LEED doesn't adequately compare energy-efficiency of buildings. See for example, Wall Street Journal, October 19, 2005, p. B6. Indeed, if energy savings are the primary goals of a project, the Energy Star system may provide a better evaluation tool (see Chapter 17).
9 For a somewhat technical description of this approach, see "Integrated Design for Small Commercial Buildings: An Idea Whose Time Has Come," by Peter Jacobs and Cathy Higgins, *HPAC Engineering*, July 2004, 44-53.
10 www.natlogic.com/Articles/BillReed_Questions.pdf

Chapter 3

11 Survey results are at: www.zoomerang.com/reports/public_report.zgi?ID=L223FGQFK35J
12 For more information, see the Building Commissioning Association website, www.bcxa.org.
13 Information on other state subsidies for high-performance buildings can be found at www.eere.energy.gov/financing and for renewable energy at www.dsireusa.org. Before relying on this information, please check with reliable sources, as these incentives are likely to change almost without notice.

Chapter 4

14 L. Matthiessen and P. Morris, *Costing Green*, July 2004, www.davislangdon.com.
15 Kats, G., et al., 2003, *The Costs and Financial Benefits of Green Buildings*, a report to over 40 California state agencies, downloadable from www.cap-e.com.
16 Steven Winter Associates, GSA LEED Cost Study, downloadable (578 pages) from the Whole Building Design Guide web site, http://www.wbdg.org/media/pdf/gsa_lcs_report.pdf.
17 *Ibid.* The authors note: "The construction cost estimates reflect a number of GSA-specific design features and project assumptions; as such, the numbers must be used with caution [and] may not be

directly transferable to other project types or building owners." (p.2)

18 Based on author's experience with multiple projects.

19 Includes outside LEED consultant and architect/engineer professional services time.

20 Source: www.greenbiz.com/news, September 13, 2004.

21 www.fsec.ucf.edu/EPAct-05.htm.

22 U.S. Department of Commerce: www.census.gov/const/C30/release.pdf, Monthly Construction Starts.

Chapter 5

23 *Making the Business Case for High-Performance Green Buildings*, U.S. Green Building Council, 2002, available from: http://www.usgbc.org/Resources/usgbc_brochures.asp. See also *Environmental Building News*, vol. 14, no. 4, April 2005, at www.buildinggreen.com.

24 See the recent "meta study" by Lawrence Berkeley National Laboratory, *The Cost-Effectiveness of Commercial-Buildings Commissioning*, that reviewed 224 studies of the benefits of building commissioning and concluded that just based on energy savings, such investments have a "payback" within five years. This study has been widely cited and can be viewed at http://eetd.lbl.gov/emills/PUBS/Cx-Costs-Benefits.html.

25 See www.portland.bizjournals.com/portland/stories/2001/12/10/daily18.html and www.ecotrust.org, respectively.

26 For further information, see the article by Jeffrey Reaves of the Portland architectural firm Group Mackenzie, in July, 2003, Environmental Design & Construction magazine, www.edcmag.com.

Chapter 6

27 *Building Design and Construction* magazine, accessible at www.bdcnetwork.com/contents/pdfs/bdc04White_Paper.pdf

28 *Building Design & Construction* magazine reported that Turner had 76 LEED APs at the end of June 2005, www.bdcmag.com, accessed July 20, 2005.

Chapter 7

29 See www.davislangdon-usa.com

30 See www.cap-e.com/ewebeditpro/items/O59F3303.ppt#1.

31 Higher education green building survey conducted by the author, March 2004.

32 See for example, Education for Sustainability—West, www.efswest.org and Engineers for a Sustainable World, www.esustainableworld.org.

33 www.lclark.edu/dept/esm/green_building.html

34 www.rsmeans.com/costdata/index.asp.

35 For a good review article, see "Environmentally-Friendly Building Strategies Slowly Make Their Way Into Medical Facilities," Nancy B. Solomon, *Architectural Record*, August 2004, 179-186.

36 See Gail Vittori, "Green and Healthy Buildings for the Healthcare Industry," 2002 conference presentation, available at www.cleanmed.org/2002/downloads.html.

37 See Penny Bonda, "Putting the Healthy Back Into Healthcare," in *Green at Work* magazine, January/February 2004, www.greenatworkmag.com.

Chapter 8

38 Based on survey of points taken by first 195 LEED certified projects.

39 See Table 8-1a.

40 Recycled content at 10% of total materials cost achieves two LEED points in Materials and Resources.

41 For certified wood, assume 4% of total materials cost is wood and certified wood is 50% of total.

42 The total market for recycled wood products is much larger, probably exceeding $400 million per year, according to industry insiders, most of it sold in home improvement centers.

43 At this point, a guess.

44 Based on carpeting at $1.50 per sq. ft., 100,000 sq. ft. per project, 1000 projects. Total market for low-VOC carpet is probably much greater.

45 Not included in total project materials cost; same for green roofs and underfloor air systems.

46 Estimated at 10 MW total, $8 million/MW installed price (total U.S. commercial/industrial installations in 2005 are likely to exceed 60 MW (source: Paul Maycock, Publisher, PV News, personal communication).

47 Estimated at 10,000 sq. ft. per system (20,000 sq. ft. average floor

plate), $15/sq. ft. incremental cost, or $150,000 per installation. Based on 53% of projects choosing a green roof at 50% coverage, or an Energy Star/high-emissivity roof at 75% coverage, and green roofs representing about 20% of the total.

48 Estimated at 200 projects, 100,000 sq. ft. per project, $6/sq. ft. premium cost for flooring system and diffusers; does not include carpet tile or other approaches to UF air systems.

49 Estimated based on 53% of projects achieving 30% water use reduction inside the building and about 80% of those using 10 waterless urinals per building at a cost of $5000.

50 Yudelson, J. 2003. *365 Questions for Your Next Green Building Project.* Portland, OR: Interface Engineering (privately published, 40 pp.)

51 For further discussion of *Sustainable Services Forums: A Green Design Tool* and "eco-charrettes," see the author's interview at: www.betterbricks.com

52 Davis Langdon Seah International, July 2004, *Costing Green: A Comprehensive Cost Database and Budgeting* Methodology, www.davislangdon-usa.com.

53 Adapted from Kevin Hydes, "Anatomy of A Green Building Project," presentation to *Engineering Green Buildings* conference, Cleveland, Ohio, July 2004

54 Survey results are viewable at: www.zoomerang.com/reports/public_report.zgi?ID=L223YU4S63VD.

55 *Building Design and Construction* Magazine. 2003 and 2004. Green Building White Paper, ed. R. Cassidy. November. Available at www.bdcmag.com.

56 Fichman, R.G. and Kemerer, C.F. 1999. *The illusory diffusion of innovation: an examination of assimilation gaps.* Information Systems Research, 10, 3, 255-275.

57 Paul Maycock, editor, *PV News*, personal communication, October 2004.

58 Moore, G. *Crossing the Chasm: Marketing and Selling High-Tech Products to Mainstream Customers.* 1999 Rev. Ed. New York: Harper Business.

59 Gladwell, M. 2000. *The Tipping Point.* Boston: Little, Brown.

60 www.californiasolarcenter.org/solareclips/2001.12/20011204-8.html, and Fairley, P., 2004. "In the U.S., Architects are Ramping Up the Design Power of Photovoltaics." *Architectural Record.* March, 161-164.

Chapter 9

60 *Environmental Building News*, Vol. 13, Number 5, May 2004. Available from www.buildinggreen.com.

61 Quotes used by permission of SMPS.

62 The *Islandwood* project (formerly the "Puget Sound Environmental Learning Center") is a 10,000 sq. ft. LEED-Gold school for environmental education on a 255-acre site, one of only five LEED Gold projects at the time in the U.S. Mithun's 2003 project, a five-story dormitory at Portland State University was LEED Silver-certified in the Fall of 2004.

Chapter 10

63 See Seth Godin, Purple Cow: Transform Your Business by Being Remarkable. 2003, available via download from www.amazon.com

64 After D.A. Aaker, *Strategic Market Management*, 6th Ed., 2001 (New York: John Wiley & Sons), p. 209.

65 M. Porter, *Competitive Strategy*, New York: Free Press, 1980.

66 M. Treacy and F. Wiersma, *The Discipline of Market Leaders*, 1995. Reading, Mass.: Addison-Wesley.

67 S. Lowe, *Marketplace Masters: How Professional Service Firms Compete to Win*, 2004. New York: Praeger Publishers.

68 E. Rogers, *Diffusion of Innovations*, New York: Free Press, 5th Ed., 2003.

69 In Oregon, for example, the state's Business Energy Tax Credit, worth 25% of the initial cost of PV systems, can be "passed through" from an institution or government agency to a for-profit tax-paying entity, on a "dollar for dollar" basis, making it available for all projects in the state.

70 After Kotler, 1997, p. 473.

Chapter 11

71 Kats, G. *et al.* 2003. *The costs and financial benefits of green buildings*: a report to California's Sustainable Building Task Force. Sacramento, CA, at p. 96.

72 This study can be found at: http://www.davislangdon-usa.com.

73 *New York Times*, April 28, 2004.

74 Cited from www.betterbricks.com, accessed October 3, 2005.

75 www.windmilldevelopment.com.

Chapter 12

76 A good case study can be found at: www.rmi.org/sitepages/pid209.php, and the Corbetts' own story can be found in their *Designing Sustainable Communities: Learning from Village Homes*, 2000, Washington, DC: Island Press.

77 See information at www.tucsonelectric.com.

78 www.earthadvantage.com

79 P.H. Ray and S.R. Anderson, 2000, *The Cultural Creatives: How 50 Million People Are Changing the World.* New York: Harmony.

80 Karen Childress, WCI Communities, personal communication. Presentation at *Greenbuild* conference, Portland, Oregon, 2004.

81 In *Metropolis* magazine www.metropolismag.com, May 1, 2004.

82 Portland *Daily Journal of Commerce* magazine, May 2004, www.djc-or.com.

83 www.unicoprop.com/property/seattle/cobb.aspx.

84 *USA Today*, March 31, 2004, www.batteryparkcity.org.

85 http://leedcasestudies.usgbc.org.

86 See article in www.buildings.com, August 24, 2004.

87 www.betterbricks.com, interview with Hazelhurst in June, 2004.

88 Portland *Daily Journal of Commerce*, September 24, 2004, www.djc-or.com.

89 *Ibid*, and personal communications.

Chapter 13

90 U.S. Department of Energy, Energy Information Administration, March 2001, *Monthly Energy Review.*

91 *Ibid.*

92 U.S. Department of Energy, Energy Information Administration, "Emissions of Greenhouse Gasses in the United States 1999."

93 U.S. EPA, 1998, "Characterization of Building-Related construction and Demolition Debris in the United States."

94 U.S. Geological Service, 1995 data.

95 Lenssen and Roodman, 1995, "Worldwatch Paper 124: A Building Revolution: How Ecology and Health Concerns are Transforming Construction," Worldwatch Institute.

96 www.metropolismag.com/cda/story.php?artid+819.

97 See, for example, Van der Ryn and Cowan, *Ecological Design*, 1995; Wilson, et al., *Green Development: Integrating Ecology and Real Estate*, Rocky Mountain Institute, 1998; Hawken, Lovins and Lovins

Natural Capitalism: Creating the Next Industrial Revolution, 2000; and Van der Ryn, *Design for Life*, 2005.

Chapter 16

98 U.S. Census Bureau press release, July 20, 2004.

99 Edward Mazria, "Designing a Better Future," *Solar Today*, November/December 2004, www.solartoday.org, at p. 26.

100 See: www.pdc.us/pdf/ura/occ/lloyd_crossing_sustainable.pdf.

101 See for example, www.natlogic.com.

102 The most recent draft is at: www.ortns.org/documents/TNSConstructionPaper-draft9.pdf.

103 McDonough, W. and M. Braungart, 2002, *Cradle to Cradle: Remaking the Way We Make Things*. New York: North Point Press.

104 Janine Benyus, 1998. *Biomimicry: Innovation Inspired by Nature*. New York: Harper Collins; and P. Hawken, et al., 2000, *Natural Capitalism: Creating the Next Industrial Revolution*. Boston: Back Bay Books.

105 See www.worldchanging.com/archives/001449.html.

Chapter 17

106 Energy Star annual report: http://www.energystar.gov/ia/news/downloads/annual_report2004.pdf.

Chapter 19

107 U.S. Census Bureau, P25-1130, Middle Series.

BIBLIOGRAPHY

Building Design & Construction Magazine. 2003 and 2004. "Progress Report on Sustainability," ed. R. Cassidy. *Building Design and Construction*, November issues. www.bdcmag.com.

CHPS, 2004, Collaborative for High-Performance Schools, www.chps.net.

Fichman, R.G. and Kemerer, C.F. 1999. The illusory diffusion of innovation: an examination of assimilation gaps. Information Systems Research, 10, 3, 255-275..

Gladwell, M. 2000. *The Tipping Point*. Little, Brown, New York.

Kats, G. et al. 2003. *The costs and financial benefits of green buildings: a report to California's Sustainable Building Task Force*. Sacramento, CA.

Kotler, P., 1997, *Marketing Management*. Ninth Edition. New York: John

Index